HUMANITY, UPLOADED
R.E. ALLEN

©2020 R.E. Allen. All rights reserved. No part of this publication may be reproduced, distributed, or transmitted in any form or by any means, including photocopying, recording, or other electronic or mechanical methods, without the prior written permission of the author, except in the case of brief quotations embodied in critical reviews and certain other noncommercial uses permitted by copyright law.

ISBN: 9781543993080 (print)
ISBN: 9781543993097 (ebook)

To my family. To those that give me freedom in thought and in action. To those that take risk after risk after risk.

PREFACE

I'm a product manager. If you're not in technology, you probably have no idea what people like me do. We're supposed to be generalists that look at different market problems and create solutions for them.

Even though we are quite a philosophical discipline, we're actually needed to prioritize and stitch things together in an organization. Because, without us, all the business units just run in different directions trying to solve various parts of a problem that either aren't important or just don't fit together. We have to think about a whole host of variables, including market conditions, business cases, core organizational assets, design, technology. Stuff like that. We do a ton of research about things we know nothing about and we look deeply within our organization. And then we bring it all together into a strategy that solves the most important thing in the industry, by making the best product, using the most valuable resources the organization has. And then we make the strategy happen by bringing all the business units together.

I've been doing this for 16 years now, and no problem seems impossible to solve. Not because I solved them all. And not because all the ones I personally solved were all solved exceptionally well either. Just because I know that all problems are solvable with the right strategy and execution. And, eventually, the right strategy and execution meet.

So, solving problems has become somewhat routine.

Even though product managers are generalists, they usually have some domain experience. Mine has been in mobility and transportation and even deeper within that it's been using the power of machine intelligence to transform the movement of people and things.

When I'm not at work I'm either enjoying a cocktail in some odd part of the world (I travel a ton) or I'm on Instagram (@hi_im.rdot). The cocktails started to be less pleasurable when plastic straws where suddenly being displaced with paper ones. It ruined my whole mojo. And then, my Instagram feed started to really suck too because all the wonderful aspirational clothing ads were suddenly replaced with sustainable environmental ploys.

Turns out, I was subject to a movement and I had no say in it. Climate change protesters were lurking their way into my free time, reaching their young hands into my life and pulling out the dirty uncomfortable truth that all of humanity is going to die if we don't take action.

The thing is, nobody really has a vision for what a resolved world would look like. So, how can they address solving the problem of saving humanity in a comprehensive way? I mean, it's not like I thought about it much either, but I just began doing what I do. I started researching. I started treating the Earth as the problem. I started looking at our socioeconomics as the market. I treated humans as the core assets to productize. I started examining all aspects of this.

In product management, they teach you to begin anything new by evaluating and comparing opportunities, not solutions. You have to do things like look at how big every addressable option is in a certain market. So, here's where I started comparing the problems that all the protesters were advocating to address with a few other things. It became a matter of atmospheric change versus atmospheric inevitability. Meaning, the problems that we face today as a result of the actions we take to damage the environment are much smaller than the problem of the Earth's natural regenerative cycle. The former can slowly wipe out a few species on Earth, but the latter can wipe us all out in one giant swoop.

In product management, once you identify the biggest problem, you begin finding the best solution. I started looking at all the options and came across transhumanism in my research. Transhumanism is the belief that humans can use science and technology to evolve beyond their current physical and mental limitations. So, I compared the practicality and likelihood of success between transhumanist ideas and green initiatives.

Transhumanism won. The rate of new technology adoption was working in our favor. It was like an upward wave that we were already riding. We just don't really realize that the wave can be used to totally solve our problem of escaping the kind of mass extinction event that could blow away the entire planet. Meanwhile, the difficulty with changing habits of the largest contributors to climate change — companies and governments — have been met with uphill battles that can't really protect us indefinitely, even if we totally transformed the way they operate. They would just leave us in this constant state of defense, vulnerable to the unknown resetting of our planet. But, transhumanism would get us excited about starting over. It would get us prepared for it. It would allow us to rid ourselves of all the crazy things like race and geographic borders and the social classes that divide us. With transhumanism, we could become a fearless group of humans that are all so incredibly different, but so deeply bonded and in need of one another's differences.

Sure, transhumanism may sound like science fiction today, but it can be made real by applying the right scientific approaches in the right global socioeconomic state. Here, both transhumanists and environmentalist groups play a role in creating the optimal future. They just need a strategy that brings them together on a common path in order to get started.

So, I started forming a realistic plan to transition the evolution of humanity. It's the kind of strategy I put together all the time. It involves a vision, a reorg to align groups, an economic analysis, and an approach to designing a technology product. That kind of stuff. Except, this time, the strategy is not about using machines to move people and things from one place to another. It is about using machines to move people from one *form* to another. One that doesn't require suffering. One that uses the best of what we have built as humans to transport us into the best existence that we have yet to experience as humans.

I know a lot of stuff in this book isn't perfect. There may be angles that I didn't see. There will certainly be typos and incongruent thoughts. As a product person, you have to get comfortable with being wrong and pissing people off. You also just have to get comfortable shipping things (i.e., your strategies, your products, stuff like that). Because, when you ship something wrong and piss people off, something really cool happens. People chime in and try to correct you and it just makes your whole product better.

So, I'm shipping my MVP (minimum viable product). If you want to provide ideas that expand on the thoughts for future editions, send me a DM on my Insta or find me on LinkedIn.

CONTENTS

PART I: VISION xiii

    Chapter 1: **The State of the Earth** 1

    Chapter 2: **The State of Humanity** 6

    Chapter 3: **Uploading Humanity Takes the Mission** 18

PART II: SOCIETY 25

    Chapter 4: **Uploading Humanity Takes Social Reorganization** 27

    Chapter 5: **Uploading Humanity Takes New Grounds** 42

    Chapter 6: **Uploading Humanity Takes Big Bets** 59

    Chapter 7: **Uploading Humanity Takes War Tactics** 70

    Chapter 8: **Note** 81

PART III: MACHINES 83

    Chapter 9: **Uploading Humanity Takes Communication With Machines** 85

    Chapter 10: **Uploading Humanity Takes Social Scoring** 100

    Chapter 11: **Uploading Humanity Takes New Decisions — Aquatic Uploaders** 110

    Chapter 12: **Uploading Humanity Takes New Decisions — Galactic Uploaders** 115

    Chapter 13: **Uploading Humanity Takes New Decisions — Mind Uploaders** 121

    Chapter 14: **Uploading Humanity Takes Physical Integration** 127

    Chapter 15: **Uploading Humanity Takes New Survival Instincts** 136

PART IV: SPIRIT 145

Chapter 16: **Uploading Humanity Takes Communication with the Universe** 147

Chapter 17: **Uploading Humanity Takes a New Belief System** 161

GLOSSARY 181

The planet is in the midst of a global environmental catastrophe. Unless we take radical steps to ensure survival, mankind will soon be extinct.

We must destroy our current society to survive.

We must transform the human material from flesh to flesh-connected-to-machine to machine.

In the process of transhumanism, we must rid ourselves not only of our flesh but of the dimensions that weaken us socially.

We must carry forth not only the best of our species but the best of its minds.

We must sacrifice ourselves to rebuild a future. New rules govern the new world, the Fourth World.

PART I: VISION

CHAPTER 1: THE STATE OF THE EARTH

"It will be a vastly different planet — as different as our current climate is from the most recent ice age."[1]

— *Gavin Schmidt*

A mass extinction

When Environmentalism became a new religion, our society got a lot less liberal. As though our social construct doesn't restrict us enough, a new prejudice formed against consuming natural resources. And now, we can't utilize the Earth freely.

Environmentalism isn't bad. It's just a bit misguided at this stage.

Environmentalism worked it's way right into corporations. That's why, in April 2019, more than 1,000 people were arrested in London. They were part of a fiery environmental activist group that began disrupting traffic, gluing themselves to trains, and smashing the windows of Shell Oil Company headquarters. It's certainly not disputable that transportation contributes to an excess of CO_2 in the atmosphere. And, CO_2 is causing global warming. So, Environmentalists aren't wrong about that.

---

1 Mosher, D. (2017, November 17). *Here's what Earth might look like in 100 years — if we're lucky*. Retrieved from https://www.weforum.org

The worst thing is, Environmentalism is reaching its way into our governments. By September 2019, a group called The Extinction Rebellion organized protests in 4,500 locations across 150 countries. They brought together about 6 million people. Students assembled in Italy. Artists took to the streets of Israel, Ireland, and the global stage to simulate the end of our species. New Yorkers stormed the streets, splashing the iconic Charging Bull statue and themselves with red paint.[2] In Angola, over 50,000 children gathered for the first time in solidarity. The group urged governments to declare the state of the Earth in a global climate and ecological emergency. Protestors demanded that the government create programs to halt biodiversity loss and reduce greenhouse gas emissions to net-zero by 2025. They insisted on creating a citizens' assembly to advise the governments on a just transition.

Trees convert CO_2 into oxygen, and there aren't enough of them to properly filter all the fuels that we emit into our atmosphere. So, Environmentalists aren't wrong to be threatened by our air filtration and ecosystem balance either.

The fear that we will turn into history's next dinosaur is backed by science too. Climate scientists say that the tropics will experience an increase of extreme-heat days. Severe drought will plague 40% of all land.

Our oceans, now teeming with life, will become acidic. The coral reef homes of fish will die and disrupt the food web that connects all living things.

Sea-levels will rise, displacing about 4 million people from the coast. This will force many to move inland. By 2100, the displaced will join a land overpopulated with 11 billion people who will occupy more regions with deadly insects. The common home will be shared with Tsetse flies transmitting fatal diseases and Black Widow Spiders with debilitating venom. The grounds we walk so freely today will see

---

2 Sometimes referred to as the Wall Street Bull, the statue symbolizes aggressive financial optimism and prosperity

a trudging of hungry, harrowing mortals. So, Environmentalists aren't wrong to be worried.

It is certainly conceivable that climate change can result in a mass extinction event. Many believe that this event is approaching very quickly — perhaps in the next 80 years.[3] And, it is indisputable that the combination of government, consumerism, and fossil fuel burning corporations are contributing to factors that make the environment more and more difficult to survive in.[4]

But what if the actions that we take to restore the environment won't make a difference in the end? What if the Earth will change, despite what corporations and governments do?

What if, together with changing *our actions*, we also build a strategy to change *ourselves* — to change our material destiny to evolve *with* the Earth?

After all, as a species, we have used the Earth to evolve to this point. We have cultivated the land to strengthen ourselves. This helped bring us to the point where our health, longevity, and reproduction rates have led to overpopulation. It is that overpopulation that has led to the degradation of the environment. To survive as a growing population, we destroyed forests for crop cultivation. 48% of deforestation is the result of subsistence agriculture, and 32% is the result of commercial agriculture.[5] We took from the Earth to feed ourselves. We are transporting more and more goods, and we are becoming more mobile.

---

3 Rothman, D. H. (2017, September 20). *Thresholds of Catastrophe in The Earth System*. Retrieved from https://sciencemag.org

4 For example, government-imposed tariffs have caused deforestation by countries unequipped to handle changing production needs. A slow down in soy imports from China to the US as a result of tariffs moved global soy production to Brazil. To make room for soy farm land, Brazil embarked on controlled rain forest fires. However, as controlled burning was not routine practice, Brazil couldn't stop the fires and it resulted in mass deforestation.

5 Naik, A. (2018, March 5). *How is Deforestation Related to Population Growth?* Retrieved from https://helpsavenature.com

The excess CO_2 from our movements, coupled with the dwindling forests, has *already* caused global atmospheric and oceanic temperatures to rise, speeding up climate change. Climate change causes a vicious cyclic effect in which the warmer oceans release more CO_2, leading to an even warmer atmosphere. This warmer atmosphere further increases ocean temperatures.[6] This is impacting the change in water *cycles* and increasing the frequency of natural disasters such as hurricanes, tornadoes, and fires that destroy towns and cities across the world. In turn, agricultural plots have already turned into barren wasteland. Meanwhile, warming waters are killing oxygen-producing ocean life and food-producing fisheries. But, there are also factors, completely uncontrollable to us humans, that can, and likely will, have a major impact on our Earth. And, this impact is not related to CO_2 emissions.

The Earth is a dynamic planet. The universe has existed for 13.8 billion years. As a civilization, we have only existed in it for ten to twelve thousand years. The Earth has *constantly* changed forms through wind, water, and ice erosion, through violent Earthquakes, and through the reconfiguration of continents by movement of the Earth's plates. And it will continue to change. Drastically.

A mass extinction event can happen if a large meteor hit us, or if volcanic eruptions clouded our skies with ash and toxic smoke. A violent solar flare can strike us directly. After all, these were some of the causes of past mass extinction events.

Mass extinction events are real and we shouldn't strive to only solve some of the reasons why they may occur. We shouldn't look to prolong our existence in its current form just a little bit longer, and it would be foolish to assume that improved environmental conditions alone will save us.

Instead, each one of us should deeply understand what it is that we need to spend our time on, as one global society, in order to survive. We should look at the next 80 years and ask ourselves, what are we doing

---

6 *The Effects of Climate Change*. (2019, September 23). Retrieved from https://climate.nasa.gov

to evolve humanity into a species capable of surviving any change to our environment, *however,* and *whenever* it may occur?

To invest in the evolution of humanity – one that is capable of withstanding any cause for a mass extinction event – it is worth exploring how we have evolved so far. It is also worth understanding what makes us human and why we became who we are today.

This understanding will allow us to consciously press our advantage to advance our species in the direction we desire.

CHAPTER 2: THE STATE OF HUMANITY

Human being /hu·man be·ing/

noun

1. *"A culture-bearing primate."*

— *Encyclopedia Britannica*

Why us?

Why have the mental capabilities of humans evolved to engineer biomedical systems that advance our health, while chimpanzees, our close living relatives, are limited to eating fruits, leaves, and other animals for nutrition? Why do we use our minds to build ships to travel into space, while chimpanzees are stuck swinging from trees? Why do humans create computer systems to instantly interact with one another from the other side of the planet, while chimpanzees can do no more than call out to one another within earshot range?

Perhaps, like every other animal in the kingdom, Chimps don't need to evolve their minds much further. After all, evolution is more about how organisms fit into an environment to survive and reproduce, rather than a matter of progression of strength and speed. Chimpanzees may be physically fitting into the changing environment just fine.[7]

---

[7] Currin. G. (2019, July 14). *Why Haven't All Primates Evolved into Humans?* Retrieved from https://www.livescience.com

Sure, the survival *rate* of Chimpanzees is declining *today*. But the rate of reproduction does not necessarily indicate the *quality* of reproduction. Chimps could very well be filtering out members of their breed to physically regenerate. They could be procreating new versions of themselves to sustain the changing environmental conditions. Those who are alive are learning to retain less water. The fibers in their muscles have distributed to become 1.5 times stronger than humans. They are evolving to withstand higher heat, and they are scavenging for food at night in preparation for something — something us humans have yet to evolve our physical form to adapt to. It is Darwinism at its finest.[8]

Meanwhile, ants, which evolved about 140 million years ago from their wasp-like ancestors in the Cretaceous period have diversified differently. They have used reproduction rates to survive, but for a unique advantage. Queens breed 1500 children. This enables colonies to bond together and float above water so that no single ant is left to drown. Their survival tactics have brought them to the point where there are more ants than humans. Like Chimps, perhaps ants have actively prepared to survive something — something us humans have yet to evolve our physical form to adapt to.

Physically, we are a weak species. Humans don't have robust fur, muscle, or breeding capabilities. We have advanced intelligence — the ability to learn from our individual experiences, to act purposefully, to solve complex problems — as the one survival trait that sets us apart. We decided that our intelligence is an all-ruling aspect — one that puts us at an advantage over the primate ancestors that we split from, and the entire population of Earth's creatures, for that matter. Yet, advanced intelligence is just our form of survival.

Why have humans evolved their intelligence differently than other living beings? After all, at the subatomic level, our minds and bodies are made up of the same neutral elements. The concept of **neutral monism** describes that these elements are neither physical, nor mental. Since

---

8 A theory of the evolution of species by natural selection advanced by Charles Darwin, Darwinism states that species develop through small, inherited variations that increase the individual's ability to compete, survive, and reproduce.

everything is made from the same "stuff," it is the way these elements engage with other elements that forms unique properties, allowing different parts of our "selves" to form and perform different functions for our survival. Humans are predisposed to working with external resources, rather than naturally evolving to complement our environment. This doesn't mean we aren't physically evolving at all. It just means that we are prone to evolving in conjunction with external materials, as opposed to other animals that evolve their physical form absent of foreign materials.

We convinced ourselves that intelligence has the power to save things other than ourselves. In doing so, we misuse the strongest function of our design. Just like other living organisms are physically altering their bodies to prepare for environmental changes, we must harness our unique strength to evolve *our* material forms. We should understand how we can use our intelligence to integrate with *new material forms*. We should seek to design new versions of humanity that can survive asteroid collisions or oceans losing oxygen or volcanoes erupting or any other global warming events.

But first, we need to understand the external factors that have allowed us to evolve very differently than all other species, so that we can revisit these factors. Once we do, we will see that the human was designed to evolve its material form with technology, right before we split ourselves up into societies with different goals.

We are an organism shaped by technology and socioeconomics

The human experience of today is a composite of technological and socioeconomic factors that manifest in our neutral material form. To say that both technology and socioeconomics create a human experience is to say that our entire *physical and mental* existence has quite literally been shaped by these factors. It is to say that technology and the way we organized to use it has created our current form.

We can use technology and socioeconomics to promote any part of our neutral elements (aka: our mind or body), while still remaining human. This is because our bodies and minds are evolving forms subject to these distinct factors. The sooner we see that our intelligence is just one part — the strongest part — of our evolved form, the sooner we can make conscious actions to traverse our entire essence from our bodies and minds into a new material form. We can shed the parts of our bodies that we don't need and merge with new materials. And we can do this and still remain human.

This notion brings into question if we are making conscious efforts, as a society, to transform ourselves using the technology we have available to us. It allows us to take into account the environmental conditions that will present themselves in the future. It allows us to evaluate how we are utilizing the Earth's materials to survive atmospheric changes. Are we truly shedding off the parts of ourselves that we don't need? Or, are we building industries to recreate the parts of us that are failing at scale? Are we set up economically to augment ourselves with new parts that can be more useful? Or, are we just using technology to make some of our parts smarter? This notion allows us to ask if we are integrating with materials that will allow us to go *even further* than our natural forms can take us today — beyond this time, beyond this Earth, beyond this dimension.

We can dissect the technology and the socioeconomics that have shaped us in the past in order to inform our choices for material integration for the future. It is worth looking at technology and socioeconomics separately, starting with how technology has shaped our existence.

Expansion of the mind

Over the past seven million years of our existence (mostly in the last two million years), the human brain has tripled in size. Because the brain takes in such a large number of calories for its size (about one-fifth of the energy consumed by the human body), any growth in size has been paid for with calories taken in or diverted from other bodily functions. Why did our brain grow? This is where technology comes in.

About 1.7 million years ago, our prehistoric ancestors developed a dazzling new technology, previously untamed by any living being — fire.

Fire led to increased survival in many respects. It allowed us to defend ourselves from predators better and to keep ourselves warm in harsh weather. Thanks to easier digestion, it changed our diet from one of tough, raw meat and plants that are difficult to chew into one that allows the body to get up to 80% more nutrients and calories. Our jaws no longer needed the bulky muscles that made speech difficult and once blocked the capacity of the brain from forming the Broca, the language-connected part of our frontal lobes. Harnessing fire profoundly changed Human anatomy by giving the brain more room to grow.

A bigger brain gave us the ability to expand our intelligence. We developed other technologies with similar effects. Tools for chopping, grinding, and mechanically breaking down food before eating it eased digestion further and diverted even more calories to the brain. Our growing brain allowed us to improve our communication. So we increased our depth of planning, problem-solving, and other more advanced cognitive functions. We perpetuated the cycle of developing more tools, becoming smarter, and developing even better tools.

Eventually, this cycle came to the modern era. Tools of our evolutionary success in brain development surround us in the form of computers, smartphones, and endless other connected devices. Integration with technology now forms a feedback loop between machines and humans, whereby humans feed data to machines in an aggregate scale, and machines learn from that data to feed individuals with information that, in turn, transforms the human brain even further.

We have access to incredible amounts of information, almost instantaneously. It has decreased our memory and attention span, yet freed our minds for deeper thinking. Information has begun to act on our brain in the same way money or food do through dopamine-producing reward systems. As a result, our neutral matter is morphing.

According to a recent study, we are growing horn-like structures on our skulls due to smartphone use.[9]

We have now arrived at the point where biotechnology has set the stage for brain and body prosthetics — allowing technology to not only affect us but to become part of us. It has begun to shape our material destiny.

Genetic engineering has the potential to catapult the human brain to become one with a machine, making us capable of producing extraordinary new feats. Yet manifestation of the new human does not depend on technology alone.

Fiefdoms

In many ways, technology and socioeconomics are tightly coupled in shaping the human. So, it is worth understanding the role of socioeconomics in this relationship.

Our prehistoric ancestors quickly learned that groups increase our chances of survival. Group harmony became essential in this effort, and every individual had a role.

This mindset carried over to the English feudal society of the 1600s. Groups operated through fiefs, or portions of land, that were controlled by nobles and knights who held working peasants on their properties. Fiefdoms were rewarded in exchange for protecting the King. People lived off their land, while towns were few and far between. There was little commerce between towns as peddlers provided just the few essential goods that peasants could not produce themselves. In a feudal society, land was the primary source of wealth, and since it was only obtained through warfare or a king's grant, peasants could really only do so much to satisfy themselves and their ruling class. They just worked to feed their bellies.

---

9 Shamard. C, M.D. (2019, June 25). *Questions raised about study linking cellphones to bone spurs in the skull.* Retrieved from https://www.nbcnews.com/health

The First Industrial Revolutions

Describing the periods that followed fiefdoms, economic and social theorist Jeremy Rifkin explained: "Before each industrial revolution, three technologies have emerged and converged forming a general-purpose technology platform. These technologies were new forms of communication, energy, and mobility. They fundamentally changed the way we managed, powered, and moved our economic life. They changed our temporal-spatial relationships: how and where we live. They allowed us to integrate in larger units. They changed our consciousness and governance."[10]

It was the introduction of capitalism in the First Industrial Revolution that began to propel us into the path we are on today.

The First Industrial Revolution was characterized by the invention of the steam engine in 1698. This new technology created opportunities for society to reorganize, which resulted in the establishment of new business strategies. The new ways of doing business consequently changed individual human behaviors. The new ways of using the revolution's technology transformed a feudal society in England into an industrial one, where individuals could compete for financial gain. It turned peddlers into merchants, and instead of capping out labor on personal consumption, individuals were able to labor for accumulation.

Just like individual engagement with fire technology transformed our physical form to then enhance our mental capabilities, economic engagement with one another and steam engine technology transformed our mental social *wiring*. It changed how we thought and transformed our methods of survival. The old feudal society taught us to labor for food to survive. But in the new capitalist society, most people depended on selling something, or at least their labor power, to earn a livelihood. They moved out of rural towns to ports and cities. They were able to trade new things in large quantities. There was no

---

10 *The Third Industrial Revolution: A Radical New Sharing Economy*. (2018, February 13). Retrieved from https://www.youtube.com/watch?v=QX3M8Ka9vUA

limit to the accumulation of wealth, and wealth could be shared with offspring (i.e., the promise of future survival).

So, mentalities changed towards an endless pursuit of accumulation for the sake of accumulation. It gave rise to a new form of competition and a new wave of individualistic thinking. It created a techno-economic framework obsessed with measuring the performance of a process, product, or service. It brought us to the point we are at now. Many of us have accumulated so much that we are just looking for better versions of what we already have, not understanding that we are operating in a model that was shaped by the survival needs we had hundreds of years ago.

The Second, Third, and Fourth Industrial Revolutions

The First Industrial Revolution transformed human psychology by providing more efficient production, lower prices of goods, and improved quality of life. Yet with it came the steam engine, which also introduced new, more powerful sources and trends of pollution.

However, it wasn't until the Second Industrial Revolution in the late 1800s, that the friction we face today between environmental sustainment and enjoyable survival was brought to light. Technological advancements initiated a new source of energy — electricity — using gas and oil. The invention of the telephone revolutionized methods of communication. Transportation advanced with the automobile, semi-truck, and train. All these inventions were made possible by the use of steel, which developed alongside an economic and industrial model based on large factories.

While the Second Industrial Revolution continued to enhance our way of life, providing even more ease in production and enjoyment with consumption, it too added exponentially more strain on the environment. The technology added — through energy production — to the millions of tons of greenhouse gasses emitted into the atmosphere each year. This contributed to the atmospheric changes that were already beginning to threaten biodiversity and the survival of the human species. While mankind was advancing alongside these damages, news broadcasters cited that never in the planet's history have CO_2 levels risen even 1% as fast as during the past 120 years. And, with education on the dangers of CO_2, a new level of fear arose with consumers. 66% of the global population reported a willingness to spend more on eco-friendly goods than equivalent alternatives. Protests and rallies formed, demanding governments to participate in actions that prevent industrial enterprises from expediting climate change.

However, the industrial revolutions were not the only drivers of the atmospheric changes. As mentioned above, overpopulation is also a contributing factor. Just like the destruction of the Earth's forests as a result of human breeding has already threatened our air filtration and ecosystem balance, the subsequent increase in air pollution as a result of more humans enjoying a better quality of life has also *already* emitted an excess of CO_2 in the atmosphere.

Be it in the Second Industrial Revolution, or today, attempting to slow down changes in the environment has a minuscule impact. Spending more on environmentally-conscious consumer goods, not wearing fast-fashion, banning plastic bags and straws, protesting to save the polar bears, trying to prevent the sea level from rising a couple of meters, and insisting on growing organic food to avoid pesticides won't save us. Our adoption as a species has fallen behind. All will not survive in their current form. The flesh will not withstand it.

We are now presented with the Fourth Industrial Revolution, which truly does unlock an opportunity to survive Earth-shattering disasters. This is because the Fourth Industrial Revolution uses technology capable of altering our physical forms to survive in *any* environmental condition. Before discussing the Fourth Industrial Revolution, one must understand the Third Industrial Revolution and the power of consumer sovereignty. Without understanding these things, the power of the Fourth Industrial Revolution will fall short on its potential.

Consumer sovereignty, also known as consumer choice, has begun to drive techno-economics on a course that will truly impact the human form. But, it didn't always sway in this direction. Different consumers tend to want different things. Consumer sovereignty guides how resources are used to satisfy their varied wants.

A conflict exists between the unlimited wants and needs of people and the limited resources available to them. With limited physical resources, we prioritize our wants, sacrificing some wants in the process. Resources are used by society to its greatest advantage, weighing all their possible uses. *Advantage* is typically related to the financial rewards achieved from that resource. Advantage is often driven (at least eventually at the end of the supply chain) by consumer willingness to pay. An aggregate group of individuals present a desire. Technology adapts to fulfill that desire. And the human species evolves with the technology.

Not solely as a result of the desire to protect the environment by spending more on eco-friendly goods, but certainly a factor of it (as individuals started to become more aware of what they were consuming), we began to see the Third Industrial Revolution emerge in the second half of the twentieth century. Accumulation of goods became less meaningful. Rather, a desire to acquire better goods faster formed a powerful group of individuals, that held the economic capital and began to sway technology.[11]

Nuclear energy emerged alongside the rise of advanced telecommunication systems, computers, and electronics made from microprocessors and transistors. These technologies gave rise to a high level of automation. Today, we see the Third Industrial Revolution manifesting itself through communication systems ignited by **Artificial Intelligence (AI)**, energy powered by renewable resources, and mobility fueled by autonomous vehicle development.[12] They gave rise to a general-purpose technology platform. Like prior revolutions that were built on new sources of communication, energy, and mobility systems (the three

---

11 Today, twenty-six people have half of the total wealth of the world. It is this group, along with the upper socioeconomic class of society that swayed power.

12 The father of AI, Alan Turing, created the technology in the Second World War to crack the 'Enigma' code that was used by German forces to send messages securely. However, AI was popularized by IBM's Deep Blue which was intended to gain wealth by serving other corporations filled with individuals driven by capitalist motives.

components that typically make up a general-purpose technology platform), the Third Industrial Revolution produced aggregated impacts. It provided an infrastructure for economic and social change.

New businesses formed around these technologies to provide upgraded products and services that respond to the select group of consumers willing to spend more. Fancy "connected" electric vehicles hit the market.[13] City infrastructure started getting "smarter" along with personal mobile devices. An information grid between everyone and everything started to take shape. The products and services that formed the grid trickled down to the groups who couldn't spend. Mobile devices became cheaper. The Internet got faster. And through the grid, more and more consumers were able to see more and more products and services. They were able to voice their opinions on the offerings. Slowly, technology began to arm consumers with a stronger voice in driving *industry* to satisfy their individual wants.

A tension formed between capitalism and collectivism. As choices among wants were a result of both resource availability and available ways of earning an income, those who earned more were better able to direct society's labor to produce goods to satisfy their needs. They owned the goods in their entirety and utilized them partially. Meanwhile, lower-waged workers were left with fewer choices of consumption.

A sharing economy took form to provide temporary access to all that was produced by society for the lower waged workers. In early, underprivileged, markets, sharing was rather routine and unorganized. In countries like Morocco, citizens stood on the side of roads, holding up their fingers to signal the desire for pooling as cabs picked them up and drivers mentally calculated cab fair. Meanwhile, in Vietnam, communities shared land. The Third Industrial Revolution allowed new sharing economies to leverage mobile technology, making sharing more accessible to new markets. It took on new forms, allowing individuals to share their homes, babysitting services, and even scooters.

---

13 Connected vehicles have cellular and/or WIFI capabilities, enabling new in-vehicle experiences.

It became more organized, and it used machine intelligence to better facilitate transactions.

Emphasis was placed on supporting collective wants as labor moved to satisfy shareable goods and services. Here, *advantage* signaled freedom, or access to consume. So, while freedom was decoupled from production of single owned assets, it was tightly bound to a likeness, or membership, of a substantial community. More and more, individuals began to see themselves not as one, but as one part of a group. They rated one another on the experience they received through sharing. They formed open social forums to discuss their sharing communities. And a search began for each individual to find a group they could consume in. It changed an individual's perception of "self."

Yet, a problem still remained: the capitalists, who pushed single asset ownership, and the collective, who pushed for sharing, both wanted what was best for themselves, and perhaps what was better for only one or two generations to come.

As we enter the Fourth Industrial Revolution, *digital, physical, and biological* systems will come together to form a new, general-purpose, technology platform. The coming general-purpose technology platform, that presents us today, fashions the Third and Fourth Industrial Revolutions to operate concurrently, as a group of individuals split the use of both the new and legacy platforms off into a dedicated track. New communication technologies will harness and collect data to create the digital assets. New forms of energy will be used to power the physical structures that house these assets. And, interestingly enough, technology used in the Third Industrial Revolution, to create new mobility systems, can be repurposed to support a biological transformation. With this new revolution, we hold a new opportunity to reset our path as a species. We hold an opportunity to rethink our path beyond the next few generations and to truly leverage the material technology, that we have created, in a way that fosters the survival of our species on an Earth threatened by atmospheric destruction.

CHAPTER 3: UPLOADING HUMANITY TAKES THE MISSION

"In order to survive in new environments, we have to change ourselves instead of changing our environment."

— Manfred Clynes

The Fourth World

Problems related to environmental sustainability are no longer the right problems to address. We now have an opportunity to use the Fourth Industrial Revolution to help us set a new path — to transform from a capitalistic society of individualistic wants into one that opens up opportunities to do something very different. Something that can institute a change like one that we have never seen before. We can transform into beings capable of surviving environments we never thought possible before.

It is the way that the Fourth Industrial Revolution technologies are used that defines the mission of a new existence that aims to reconstruct our material destiny — the **Fourth World**.

The Fourth World does not bear the political and economic beliefs that divided former worlds. The terms First, Second, and Third World arose when tensions between Western capitalism and Soviet socialism materialized in the Cold War. French demographer, Alfred Sauvy, wrote of "Three worlds, one planet" in L'Observateur magazine, labeling the classification system that we use today. The First World consisted of the U.S., Western Europe, and their allies. The Second World was the Communist Bloc: the Soviet Union, China, Cuba, and their allies. Countries not belonging to either group, many of which were former colonies of the First or Second Worlds and very poor, were dubbed the Third World.

Members of the coming Fourth World will not be divided by the socioeconomics that separate us today. Rather, they will be bonded by the promise new technologies hold. They will rise up in an economic transition, taking advantage of an industrial revolution, pushing us forward on a deliberate path. They will change the way we think and interact with one another. They will build a system that doesn't just have the power to change how we will experience life tomorrow, but it will have the power to transform us into beings that will be made from, and eventually by, machines — to Upload humanity.

A distinguishable feature of the Fourth Industrial Revolution is that, unlike the previous revolutions, it doesn't just have the capabilities to change how we produce or consume. Rather, it has the power to give us control over the neutral elements of our makeup. From here, we see the vision and mission of the Fourth World:

- *The vision is to exist in an unobstructed transcendence of self over space and time.*
- *The mission is to use the Fourth Industrial Revolution to evolve human beings into a new material form that can survive any environmental change.*

Early members of the Fourth World will create a new socioeconomic structure that paves the way for new technology. With new technology, a path toward Uploading humanity can be forged for later members of the Fourth World. In computer terms, Uploading simply refers to a concept of transferring a file from a computer to another system. However, in the context of the Fourth World, it means so much more. It's about the relocation of the human consciousness into a material content that can survive a regenerating planet by melding our current human composition with machines. It is about forking our species off into three tracks:

- Galactic Uploading
- Aquatic Uploading
- Mind Uploading

The socioeconomic and technological changes needed for Uploading to manifest itself will not enable incremental enjoyment; rather, they will provide a euphoric existence. The changes will take us down a path that won't just make us slightly happier. It will be engineered to affect the parts of our brain that can make us extremely happy. This is because it is founded on a science that influences our neurological triggers. Through the path of integration with machines, we can choose how we will evolve our neurological makeup and produce one of the three new species.

Galactic & Aquatic Uploaders

There will be a transition period where humans will interact more closely with machines before we can become one with them, in a world shaped by that condition. During this period, as machines enter the human flesh, the separation of the two forms will be apparent. Machines will be visible attachments. They will be integrated with us as biomedical systems to create cyborgs. While on Earth, cyborgs will have the benefit of heightened senses, higher-performance prosthetic parts, and exoskeleton shells. Their machine parts will be able to connect to the Internet to provide communication and feedback. Eventually, the integration of human and machine will appear seamless.

If not for curiosity alone, then for hope in the unknown, a group of cyborgs will stay on Earth to survive violent atmospheric changes. These few will form one branch of humanity called the Galactic Uploaders. Galactic Uploaders will advance to alter the human body's temperature, gestation, hydration, metabolism of oxygen, and reliance on exposure to sunlight.[14]

Meanwhile, other cyborgs will stay on Earth in sea bases on an Ocean Atlantis (i.e., stations built under the sea) to survive and guide the planet's regeneration. They will form a separate branch of humanity called the Aquatic Uploaders. Aquatic Uploaders will also benefit from alterations of the human body. However, their material human form will be different than the Galactic Uploaders, as they will need to survive in different conditions. They may develop tougher exteriors to handle underwater pressures or mechanisms to metabolize water into oxygen.

Over time, the conditions that caused a mass extinction event to occur on Earth will settle. However, the transformed Galactic and Aquatic Uploaders will become more capable of surviving in their new environments and will find less value in returning to the Earth's land.

Mind Uploading & brain emulation

Both Galactic and Aquatic Uploaders will depend on interaction with a fraction of new humans who will have fully transferred their consciousness into machines. This fraction of humans will be called the Mind Uploaders. They will serve as time capsules, left behind on the Earth's land, protected by durable materials and infrastructure for survival.

Inside our heads is the most complex arrangement of matter in the known universe. Yet we are learning more about it. As we advance our understanding of the brain, we advance our ability to create tools that specifically focus on mapping more of its activity.

Mind Uploading will evolve our material form by using the brain's mapping to transfer human consciousness to machines. Human consciousness transfer will capture the neural connections of a brain and store them on a computer that will use Artificial Intelligence to

---

14 Perhaps we can travel to other galaxies in the future, but the focus of this book is what can be achieved with initial Uploading. The technology for intergalactic travel is still limited by time and resources .

simulate the brain's information processing ability. It will respond in essentially the same way as the original brain and experience a conscious mind.

To believe this is possible, one must believe that consciousness is just a physical organization of matter and that machines can capture this organization. Here is where the concept of neutral monism comes into full validation. We will truly see that our neutral material elements can be transferred and reproduced to create more durable neutral material elements.

This is the ultimate end goal for the Fourth World. In this form, human consciousness can exist for eternity. And once we have achieved this, we can transfer *"packets"* of our *"states"* within an ecosystem that is able to receive and process the packets fully. In Information Technology terms, a packet is a collection of data sent between computers across a network, and a state is a memory of that packet. As we become more connected in the Fourth World, packets will represent groups of neuron connections, and states will represent our memories. Both these forms of data will be transmitted in their exact form between humans.

AGI: The rise of Distinct Intelligent Machines

A group of distinct machines will engage with the network that bridges communication between all fractions of Uploaders. They will exist to power, maintain, and protect the Uploaders. The machines will be built with **Artificial General Intelligence (AGI)**. AGI is the intelligence of a machine that can successfully perform any intellectual *task* that mirrors a human's ability today. These machines are being programmed very differently from the machine parts that will integrate with the Uploaders. They will be programmed to optimize certain, often isolated, outcomes. They will lack the ability to make a moral judgment.

The **Distinct Intelligent Machines**, simply defined as machines that will support Uploaders and possess AGI, will, however, capture our current human sentiment through a feedback loop. They will take in our aggregated data and be trained to act on the basis of this data. Their actions will support humans based on our direction.

Constructing the Fourth World of Uploaders and Distinct Intelligent Machines can be the most beautiful and liberating thing that our current form of humanity has yet to experience. We can take mother nature by the horns through a will of consciousness and program machines to help us transform our species.

But first, we must understand how to break the conditions that weaken us today. It will involve us ridding ourselves of the socio-economic imbalance that exists in society. It will involve investing in big bets. It will involve engaging in conscious communication with machines, adopting a belief in our true neutral material form, and understanding the unimportance of the cycling thoughts that consume us as individuals.

Regulations will need to support Uploading. Consumers will need to demand Uploading over any other good or service. Infrastructure will need to be present for Uploading to develop.

None of these factors can mashup, however, unless individuals are able to look outside of themselves.

We can start by understanding how to reorganize the factors that drive us to act the way we do today.

PART II: SOCIETY

CHAPTER 4: UPLOADING HUMANITY TAKES *SOCIAL REORGANIZATION*

"There is only one class in the community that thinks more about money than the rich, and that is the poor. The poor can think of nothing else."

— Oscar Wilde

Mapa Sapa

In the remote northwest mountains of Sapa, Vietnam, a woman by the name of Mama Sapa wakes up every morning to feed her cows. She makes breakfast for her husband and five children over a little fire. Then she goes to seed her rice fields and harvest her vegetables.

"When I was 14, I was kidnapped," Mama Sapa explains. "I was blinded and taken to a house. I sat there for a week. Food was brought to the door. Presents were brought to the door. At the end of the week, I was let out of the house. I had to make a decision."

Mama Sapa's decision was to marry the man that kidnapped her, sight unseen, or return to her family home down the road.

"I chose to marry him," Mama Sapa said, "and I am very happy."

Mama Sapa *is* happy. Her whole tribe of Hmong people are happy. Kidnapping young girls is a courting tradition among the Hmong tribe. It is a symbol of their outlook on life: Take what you can, nurture it, and return it back to the world. They don't look at being kidnapped as a threat either. This is because there is trust in their community.

When Mama Sapa is done with the rice fields, she returns to her 600 square foot home. She makes up the cots that her family sleeps on. She goes into her prayer room and gives thanks to the dead.

At night, her fellow Hmong come over to join random groups of international travelers that made their way through the mountains on a hike. They sit with Mama Sapa's family, drinking rice wine and eating her harvests.

As a tribe, the Hmong people don't depend on anyone outside their community for happiness. They feel freedom in meeting their own needs, themselves.

At night, Mama Sapa goes to bed thinking about the future of humanity.

The real problems in our world

Individuals cannot focus on advancing humanity until their present, basic needs have been met. But, most are never satisfied. And society is to blame for that. We primed our social construct for dissatisfaction, ourselves.

Early on, our primate ancestors learned that they could maximize their chances of survival by living in groups. When groups formed, activity needed to be regulated. A natural regulation formed in the paleo-limbic brain. It started to define one's position towards the group. The thinking organized weakest members on the outskirts of the group to be killed by predators first. The ones in the middle were seen as most valuable. Lonely, but powerful, they garnered attacks on other groups, but they were still trusted to make decisions for their unit.

In modern times, we put governments and corporations right in the middle. As we saw from the Industrial Revolutions, we programmed our brains to have these two bodies protect and nurture us. To this day, it keeps us selfish with their time and efforts. Moreover, it confines us to standards that center on disciplined behavior and likeness to peers.

We now see tensions between governments and constituents, failures in modern companies to perform under pressure, and a widening economic gap that leaves many citizens unfulfilled. On the lower end, individuals crave basic needs such as food and water. On the upper end, individuals exist in a tiny bubble and feel a sense of emptiness from being unlike others. These present market conditions are prime for disruption.

Fragile times can push society in any direction. Historically, society often took a short-term focus to repair itself. This time, we must be deliberate about taking a long-term approach by laying out a pathway for the new material form of humanity. And we can only do this if society reorganizes the way it operates.

The reorganization will entail forming a new relationship with both government and enterprise to liberate individuals from a dependency on these parties for self-sustainment. It will change what we demand from these two parties. In this uprising, we can free ourselves from the tyranny of socioeconomic divide, and receive immediate and powerful gratification from within. Only then can we send these bodies off to make the best use of our land and protect the future of humanity. Only then can we work, as a joint society, on advancing the Fourth World to create Uploaders.

Government & a shift in subordinate wants

It is worth exploring the strong relationship between the individual and government. Governments exist, today, to rule the subordinate ends of consumer satisfaction. Their role of fulfilling secondary consumer desires typically manifests through provisioning economic goods such as roads, armed forces, and protection.

Most modern governments focus on **paternalism under collectivism**, the idea that governments can act on behalf of a person's freedom to benefit constituents. In economic terms, this often translates into programs that provide goods and services to consumers who may not have access to them. For example, it can involve taxing the upper economic class of society to provide food stamps for the lower economic class. Paternalism under collectivism can also include securing resources that the public may not desire. It can provide those resources to select groups or use them to protect the people. Cranes that may seem unnecessary can be obtained and provided to select businesses to build up a city's infrastructure or weapons can be procured to protect their constituents, who may not agree with warfare. Generally speaking, paternalism under collectivism is done with positive intent. It was designed to secure the current and future good of the people.

In exchange for meeting the subordinate needs of the constituents, governments require a sort of order. Individuals are required to be unified so that governments can represent the largest sets of needs. Individuals are required to be orderly so that governments can focus on advancing the benefit of the nation. This order penetrates the psyche of each individual.

In a perfect world, governments could be trusted to either procure materials that help the evolution of humans or to meet constituents' needs so that the public can take part in advancing humanity. Yet, governments often use the Earth's assets for concealed benefits. They have proven to seek territorial control to further their personal financial or power-driven agendas rather than human advancement. This often frustrates people who believe that governments don't always act in the best interest of the nation. In this vein, people who dislike conformity and obedience to begin with, struggle with the exchange in the relationship.

Take Cameroon, for example. In the era following the end of the First World War, the Paris Peace Conference founded an intergovernmental organization called the League of Nations. The League of Nations appointed France and England as joint trustees of the German colony of Kamerun. The people struggled for their freedom, and by 1960 the French Cameroon gained its independence. In this time, a reunification referendum formed in British Cameroon that caused a divide between French and English speakers. The English speakers wanted to become their own Ambazonian nation, separating from the French altogether. These separatists waged decades of battles to fight for what they wanted. The first Ambazonian demonstration drew international attention in 2016 when the French government initiated a 93-day blackout of Internet services in the English-speaking territory following a peaceful protest of separatist teachers, students, and lawyers.[15]

The president of Cameroon, Paul Biya (who has been in power since 1982), had good reason for using aggressive tactics to keep the nation united. Less than 5% of Cameroonians had a bank account. While the nation is poor, the land is rich for cultivating crops such as sugar and nicotine. However, those are steadily losing popularity. The Ambazonian regions, however, contribute greatly to the nation's economy with palm oil, an asset Biya doesn't want to lose control of, seeing it as an opportunity to drive future wealth to a territory he currently owns. Palm oil can be used for a range of applications, from diesel alternatives to edible consumer products. Yet, these benefits only provide short term gains for Biya's regime.[16]

Following the Internet blackout, tensions rose further as the Cameroonian military began beating citizens, torching homes, and

---

15 Searcey. D. (2018, June 28). *As Cameroon English Speakers Fight to Break Away, Violence Mounts.* Retrieved from https://www.nytimes.com

16 It is arguable that the potential benefits of palm oil are why the government of Cameroon is not giving in to separatists who control its land. One could also argue that such great extremes are not necessary to preserve palm oil. Globally, today, palm oil is perceived as unhealthy and its production a danger to the environment. Nestle has excluded palm oil from its production, while Casino has banned it from all its food products for health considerations. Meanwhile, activists protested against deforestation for palm oil production. Still, there are many benefits of palm oil, including its antioxidant properties and a link to the prevention of health diseases. While 90% of it can be used in a wide range of food products, the remaining 10% can be consumed by a variety of other industries, from biodiesel to cosmetics to pharmaceuticals. The production of palm oil requires far less land, fertilizers, pesticides, water, and fuel than the production of other vegetable oils such as soybean oil, allowing for three times more oil delivery per unit. Future vegetable oil demand for the projected 2050 human population of 9.2 billion would be 240 million tons — about 40% more than today's demand. Responding to the additional demand would require either 95 million hectares of land devoted to soybean production or about 19 million hectares devoted to palm oil production.

killing unarmed civilians suspected of being separatists. Ambazonian children were banned from schools, while the separatists themselves fought back by burning markets, launching attacks from civilian bases, beheading soldiers, and kidnapping Francophile school children. Unrest remains to this day from a people looking for freedom. It is no different than the situation of a people once looking to free themselves from the German colony of Kamerun or the occupancy that followed by France and England.

The situation in Cameroon is just one of many, where people rebel against a self-serving government. Violent uprisings, lootings, and rebellions are exhibited on every continent as a response to corruption, opposing ideals, and ultimately manifestations of a people not getting their subordinate needs met by a body that holds responsibility to meet them.

What the situation in Cameroon serves is an example of a failure in focus on our future as a result of a loop that leaves us stuck, relying on governments. Governments use resources to advance short-term agendas. The ulterior motives of government officials leave people struggling to conform. People react by fighting for their current state freedom. They look to new governments for leadership that, one day, a new people will likely rebel against. A tension is formed that distracts from the purpose of the Fourth World, as long term survival of humanity is overshadowed.

As such, we cannot put the government, in its current state, in charge of our needs or the procurement of materials that will advance our material form as humans.

The fall of the modern enterprise

In many privileged nations, when consumers are dissatisfied with governments, they lean more on enterprises to solve their problems. This has created a new set of issues, as it put more stress on companies than they are prepared to solve.

As discussed in the previous chapter, the Third Industrial Revolution created a general-purpose platform that removed barriers to entry in many industries. Modern technologies allowed enterprises to step into the role of the government by providing access to secondary wants. Suddenly, every company formed or transformed into a technology company. Emerging startups made transportation more flexible with apps. Companies sprouted up to democratize education by providing easy access to lessons online. The general-purpose platform also helped create economic opportunities for individuals through platforms that enabled any individual to sell goods. But, most importantly, enterprises using modern technologies were able to deliver secondary needs with more convenience than governments.

It was the same set of modern technologies that caused market pressures to rise. The moment consumers got a taste of instant access to goods and services, they began to demand them. Consumers also began wanting higher quality. Enterprises had to compete, and many started offering goods and services instantly, either at lower costs *or* at a higher value. As consumers began to demand it all, many enterprises struggled to meet consumer-driven demands for instant gratification, lower prices, *and* better-quality products. Those who couldn't compete went out of business.

There were, of course, a few technology companies that were able to personalize at scale to serve the entire consumer market. For example, Amazon built both a set of capabilities and a sizable customer base, allowing it to produce any and every consumer product based on streamlined feedback. Coupled with data about popular product attributes gleaned from supplier partners (like the smell of lavender or the weave of pillow cushions), Amazon uses large datasets oriented around product ratings and purchase trends to serve consumers with quality products. They launched everything from Amazon-produced soap to Amazon-produced furniture. This strategy has substantial impacts on growing the business — each new Amazon-produced product directly grew its revenue. From there, Amazon was able to go deeper into the value chain to respond to consumer's desire for instant access. The company invested in leveraging a massive distribution network, allowing them to control the speed of their delivery.

It is conceivable that strategies employed by enterprises, such as Amazon, would allow for the kind of flexibility vital to the Fourth World. After all, they can leverage technology to meet the primary wants of consumers and could very well use technology to meet the secondary wants of consumers in ways that governments cannot. Companies like Amazon could use efficiency gains to fork off divisions that would allow them to begin harnessing material necessary for the advancement of our species. They could take a forward-looking approach to save humanity from the catastrophic environmental conditions that plague us while the machines they build just "work" to meet the needs of the people.

Yet, if we examine the inner workings of corporations today, we see them crumbling from within. Self-interest and personal goals create leaders looking to advance their personal agendas. It is no secret that some Amazon employees are not happy. It is no secret that they represent a group of technologists that leave their company after a year.[17] In most companies, a high employee turnover rate often results in the slowing down of company progress. More importantly, this kind of turnover signals that even the most robust modern-day enterprises aren't capable of creating the Fourth World in their current construct because they don't have the right structure for employees to fulfill a deeper, life-long purpose. Just like the tension that is formed between some governments and their constituents that seek freedom, tension is formed between technology-based employers and employees that distracts from working towards humanity's survival.

While it is hard to imagine darling companies like Amazon, Facebook, Uber, or even Google failing, it is possible that they will follow in the government's footsteps by changing the role they play in our daily lives. The inefficiency of labor and the lack of satisfaction with employees will render them unable to fulfill every consumer's need. After all, only 12% of the Fortune 500 companies in 1955 were still on the list 62 years later, and the other 88% have either gone bankrupt,

---

17 Johnson, T. (2018, June 29). *The Real Problem With Tech Professionals: High Turnover.* Retrieved from https://www.forbes.com

merged with another firm, or have fallen from the top of the list.[18] So, it is conceivable, if not inevitable, that giants of today will lose market dominance at some point. With market conditions such as these, we cannot put modern enterprises, in their current state, in charge of our happiness or the Fourth World transition.

A third problem: labor inequality

As governments in some nations fail to serve their people and corporate technology giants in other nations fall, we can look at economic trends from a macro-scale. Here we see conditions hinged on our state of labor affairs.

If we look around, we see machines aiding in our production of all goods. We see a rise in knowledge workers entering the workforce to make use of the machines. Standard economic theory states that improved productivity is a result of better machines and better workers. In actuality, this only impacts 17% of society. Today, very few individuals participate in the churn of high paying corporate jobs. Technology is only automating the production of products that benefit a small group. GDP growth is slowing globally.

Previous industrial revolutions have led us to a state in which 40% of the population is better off than they were before these industrial revolutions, while 40% of the population is worse off, making less than $2 a day. This imbalance causes inefficiency in our labor system.

As a collective, individuals are not participating equally in the economy. The ratio of high-value contributors to low-value contributors leaves some with an overabundance of resources, while others find themselves sorely lacking — a disproportionate amount of labor forces some wants to be chosen over others. Moreover, those that overproduce also overconsume.

There are three economic principles to consider here:

---

18 According to a study by Innosight, in 1965 the average company stayed on the S&P 500 index for thirty-three years. By 1990, this decreased to twenty years, and by 2026, they expect the tenure to drop to fourteen years. Retrieved from http://www.aei.org

1. Substitution

The economic principle of substitution states that when there are many goods and services in the market, the upper limits of consumer needs are acquired through equally desirable substitutes. This means that individuals will purchase products at lower price-points than the highest value alternatives. There will be excess products in the market, and the labor to produce unused products is wasted.

2. Marginal Product of Homogeneous Labor

This theory says that we must all produce units of labor equal to the value of an optimal marginal product; otherwise, the movement of labor will take place until it is equal.

3. Ceteris paribus

Ceteris paribus states that the maximum average production per head should equal the total of goods produced for the numbers drawing upon the total for consumption. In other words, we should only produce enough to be consumed.[19]

These three principles, taken together, show that the needs of the wealthy are usually met first by a large group of producers because it is expensive to create the first versions of a product. Eventually, the product becomes cheaper, and more people purchase the cheaper version. Yet the wealthy want even more goods, leaving a few people's needs met and creating gaps in producing goods for the other classes. The opportunity to further marginalize products, to provide access to the lower-income masses, is lost. Labor continues to shift gears, chasing the unmet needs of society as a whole.

---

19 Little, L.T. (1956). *Economics for Students.* London, UK. Jordan & Sons Limited.

One could argue that there are factors outside of social needs that driver labor. While this is true today, if there is a need, there is a labor opportunity to meet it. The problem is, a single mechanism for understanding needs isn't in place, but this is slowly forming, and we will discuss it in depth in the next chapter. For now, it is important to understand that unless each individual in society is producing the same amount of value to meet everyone's wants equally, those that overproduce or underproduce skew the system, leaving society unfulfilled. We remain in yet another infinite loop, widening the gap in classes with each cycle.

With conditions that create widening social gaps, it is hard to imagine a world where all individuals deliberately focus on transforming humanity into a new material form. Even if we examine Maslow's Hierarchy of needs to understand the motivational theory in all our actions, we will see that at both ends of the spectrum, individuals aren't happy. At best, individuals who have satisfied their deficiency needs will seek to achieve self-realization.[20] But, here they suffer because, as social animals, they are left outside a group. There aren't many people like them, so they strive to fill a hole. Often, that hole is filled by excessive purchase. In reality, they can't fill the hole by anything other than belonging to a community. In the worst case, and the case representing most individuals, when deficiency needs are not met, motivation to fulfill such needs becomes stronger the longer they are denied. As such, we cannot demand that the laborers, in their current state, be responsible for creating the Fourth World.

Creative destruction

As our social construct widens the gap in classes, as enterprises fail to fulfill the broad primary wants of consumers, and as governments fail to meet the secondary wants of their constituents, a shift can, and will, occur towards a new global regime.

---

20 Deficiency needs are physiological, safety, love, belonging, and esteem.

The German Marxist Sociologist Werner Sombart first framed the destruction of any given economy in his book, *War and Capitalism*, written in 1913. Dubbing it "creative destruction," he describes that, be it through war or economic crisis, capitalism destroys previous economic conditions by devaluing existing wealth to clear the ground for the creation of new wealth.

In his book, *Capitalism, Socialism and Democracy*, written in 1942, Economist Joseph Schumpeter examined creative destruction further. Schumpeter described that the eventual demise of our economic system will result in processes such as downsizing to increase efficiency and dynamism.

If we examine the global tensions today, we see an opportunity for the Fourth World to establish a new operating model. Members who create it will take advantage of the crumbling inner workings of our society. They will come to a realization bounded by drastic change and a limitless future. They will lead the resolution to it, finding ways to suppress those who are now superior, by marginalizing them with a newer, better way to enact change with a new system. They will enable individuals to get their needs met so that we can focus on restructuring humanity.

A social reorganization is slowly taking shape today. We can look at two examples how creative destruction fuels it:

1. Rwanda, war, and devaluing wealth.

In 1916, Belgium won control of Rwanda. The Belgians separated the Rwandese people into two groups based on the size of their noses and the number of cows they had. They called the people Tutsis and Hutus. The Tutsis had longer noses and more than 10 cows. They represented the wealthy society. The Hutus had shorter noses and less than 10 cows, representing lower society. In 1961, Rwanda won its independence, but the people remained separated by the two classifications. In 1994, a civil war broke out. The Hutus took a flawed approach to reclaim a new wealth. Their revolt resulted in a mass genocide that left 1 million Tutis dead. The international community turned its back on the killings.

Paul Kagame, a Tutsi residing in Uganda, rose to prominence in the Rwandan Patriotic Front and entered Rwanda to stop the civil war. Now in charge of Rwanda, Kagame was left with a tough decision: how to repair a broken and separated nation. He could have kept his people divided by their economic and physical traits. Instead, he forced Hutus and Tutsis to unite as one people. On the last Sunday of every month, the people are all forced to come together and do community service. It is a measure to reinforce the unity and equality. If one asks a Rwandese today whether they are Hutu or Tusi, they will answer, "I'm Rwandese." They will tell you, "I am truly happy being part of one community." The construct that once separated them is now creatively destroyed. As a nation, they can now leapfrog poverty to create efficiencies that benefit the future of humanity together. One such example is Rwanda's launch of a rocket that gives Internet to all schools.[21]

2. Uber, scooters, and dynamism.

Uber tore down barriers to entry for confluent groups to serve one another with shareable assets. High-salaried workers sit in the backseat of hourly workers' vehicles. Daily, these two groups talk and share stories. Yet when individuals *across* socioeconomic classes engage with each other, they don't like what

---

21 While Kagame was made President of Rwanda, and this symbolizes a form of government that goes against the Fourth World construct, what Kagame did was clear the ground for a people to rise up together and further devalue existing wealth.

they see. Together, they aim to create a dynamic new system.

While Uber broke a social boundary, the company's need to generate wealth put them in a position to be devalued. Competing price pressure has resulted in Uber lowering driver wages to gain ridership. Drivers, already struggling with the desire to be like the other half, protest Uber to raise minimum wages constantly. Yet, there is a flaw with these protests: They rely on one party to provide income.

Meanwhile, Uber riders are frustrated with Uber's internal practices. Sperate bathrooms for employees and drivers, sexual harassment, and misuse of consumer data continue to upset a public looking for equality.[22] This group of riders is represented by technologists that really *do* care about serving the other half.

Enter scooters. This new type of shareable asset has flooded the streets of major cities as an alternative transportation mode. Scooters are much cheaper than vehicles. For an Uber driver accustomed to sharing a transportation asset, scooters could very well be an additional revenue source that augments their rideshare income. While scooters are owned assets tied to large-scale platforms today, it is only a matter of time before these assets will be owned by individuals and put on a new peer-to-peer platform for sharing. Individuals will be responsible for maintaining and rebalancing the scooter to locations with high demand. The driver produces based on the service provided, rather than being bound to a single organization. It will disrupt the wealth of Uber. And it very

---

[22] Baker, S. (2019, December 5). *Alexandria Ocasio-Cortez slammed Uber's 'classism' for having separate bathrooms for drivers and employees in one of its offices.* Retrieved from https://www.businessinsider.com

well can be developed by an Uber rider, simply to help the drivers they interact with every day.

The Fourth World will build its foundation from unification and the Gig economy, which has already shifted our social paradigm to one where consumers look to fulfill one another's needs across economic boundaries. **Unitism** is a socioeconomic model that allows individuals to produce according to one's ability and consume according to one's contribution, absent of any public or private ownership of goods.

The new system is built on our innate wiring to be part of a community. It will take the form of a distributed contribution system. Individuals can achieve earnings from many parties through a system that effectively evaluates and coordinates meeting the multiple needs of their communities. It is a system that is, however, predicated on a sheer care for one another.

Here is where a shift will begin by relieving tensions and enabling us to create Uploaders. By wanting to serve one another, we can imagine a world where individuals alleviate pressure on both the government and enterprises to satisfy their needs. It represents a loss of power for what these two bodies serve, yet opens up the opportunity for them to reorganize. The new structure can shift their focus from supporting the people's well being and economy to supporting the infrastructure of Fourth Worlders who will acquire and repurpose materials that advance Uploaders.

There is a notion of technology, physical location, and community in this reorganization.

CHAPTER 5: UPLOADING HUMANITY TAKES *NEW GROUNDS*

> *"Surely you are not going to silence me. I am a woman who came from the cotton fields of the South. From there I got promoted to the washtub. From there I was promoted to the kitchen cook. And from there I promoted myself into the business of manufacturing hair goods and preparations. I have built my own factory on my own ground."*
>
> — Madam C.J. Walker

Kanye West made new grounds

Kanye West was moved to break boundaries between socioeconomic classes. It was the impetus behind funding the futuristic housing community that aimed to provide an egalitarian lifestyle. The project sat on 300 acres in Calabasas, California. Houses were shaped in huge domes. Kayne had intentions of including underground living spaces right before his project was shut down by the L.A. County Department of Public Works for violating building codes. Since then, Kanye invested in Cody, Wyoming, where he believed the government would be more sympathetic to his humanitarian construction experiments.[23]

---

[23] Kanye now owns 4,500 acres of Wyoming's Equality State, and is leasing another 4,500 acres from the federal government.

Kanye West is joined by a group of tech companies in San Francisco who are banning together to bridge the economic divide. Airbnb and Twilio gave $2.7 million to aid homelessness initiatives that provide temporary housing and job-placement services for San Francisco's homeless young adults.[24] Meanwhile, Apple donated $2.5 billion to the city for affordable housing to keep service workers and teachers in the community.[25]

While Kanye West and the tech companies represent the notion of generosity that will propel creative destruction, they face a problem: They are trying to fund a system that has already proven to be broken. It still relies on a few central bodies to meet the needs of the people. It is constrained by territorial boundaries that were designed around producing capitol. It feeds right into a system that is crippled by individuals bound to navigating long windy roads, just to get to a destination where they can earn a livelihood.

Other groups, such as those founded on communism and socialism tried to organize systems that would provide more equality as well. Communism and socialism are similar in their belief that "the people" should own economic production and resources should be allocated back to the people by government bodies. With both theories, market prices cannot develop because no one can buy or sell resources. The primary difference in these philosophies is that communists believe economic resources should be given to individuals based on the individual's needs, while socialists believe that all resources should be provided to people based on what they contribute to society. There are many reasons why these systems failed in the past. A primary reason is that these systems rely on a central planning unit that must respond to the most urgent needs of society. Evaluation of needs is complicated, and the central unit often gets it wrong. A rarely mentioned second reason for the failure of these systems is that they

---

24 Kawamoto, D. (2019, Jan 28). *Why Airbnb and Twilio are funneling millions into this homelessness partnership.* Retrieved from https://www.bizjournals.com/

25 Perez, S. (2019, November 4). *Apple commits $2.5 billion to address California's housing crisis.* Retrieved from https://techcrunch.com

are imposed in physical territories that aren't turley organized for individuals' production.[26]

Then there's the idea of post-capitalism. Post-capitalism refers to a set of proposals for a new economic system to replace capitalism, which advocates for free markets controlled by private owners. It replaces our monetary system. Here, as a result of the rise of income inequality and repeating cycles of boom and bust caused by public markets, society divides into groups of knowledge workers or service workers. Ordinary citizens become owners of organizations, and every organization will be highly specialized in their particular field. Citizen committees replace the government role by deciding how to allocate resources. Opponents of post-capitalism argue that it may fail because it is too utopian. It ignores the ingrained role of money in our daily lives; money that fuels our motivation, money that provides us with choices, and money that can't be dissolved by bartering.

For the first time in our history, we have the technology to make evaluation of value possible. The technology pinches our psyche at a point that incetivies us to produce quality for others. It reduces the need for choices because we will have optimal goods available for us. We are slowly organizing in communities that align with our interests, wiring our brains to serve one another on an even deeper level. It reduces the desire to acquire more money because our basic needs are met by communities that give us a sense of belonging without money.

Through the new technology and physical reorganization (if strategic and deliberate), individuals can be left to meet their needs themselves. Only then can investments shift to fund the future of humanity. Since technology will be the factor that makes the evaluation of value possible, absent of any public or institutional ownership of goods, it is worth examining emerging technological developments first.

---

26 Greaves, B. (1991, March 1). *Why Communism Failed*. Retrieved from https://fee.org

The rise of distributed networks

Individuals are organically collocating within communities that align with their interests. There, they share assets. Groups of YouTubers live in one home to share recording gear. Scientologists flock to neighborhoods to be near their Churches. Waves of travelers bond together to engage in remote work projects. Their social organization represents a survival mechanism against the impediments of governments, enterprises, and inequality. New technologies will enable more and more of these communities to form by organizing the transactions between members. Technology will gather the necessary information within new cities to perform a form of economic calculation.

While most of our data transactions today go through code controlled by a few major companies, such as Google, Apple, and Facebook, a decentralized web is emerging. The decentralized web offers opportunities to chat, share images, buy and sell goods, or trade currencies more privately.

It is known that pornography fueled the popularity of the Internet, changing how we consume information. However, it may be less evident how black-market activities are currently playing an active role in transforming the Internet into a catalyst for individuals to fulfill one another's needs.

In the early days of the Internet, from the 1990s to 2011, digital black-market activity operated on an unencrypted network. Opportunists found it perfect for facilitating better communication, marketing, and payment methods. While many participants on the Internet used pseudonyms, the network didn't actually conceal online activity. Sellers of physical black-market goods such as drugs, illegal weapons, and false identification papers, who found marketing and digital payment transactions too dangerous, operated in close-knit networks by word of mouth and accepted physical payments. Call-girl rings used a blend of both digital and offline means, marketing their services and accepting digital payments online, but still operating in close-knit networks. It was totally inefficient.

In 2011, Bitcoin fused the development of sites such as "Silk Road," a darknet market hosted on anonymous overlay networks called Tor.

Sites concealed identity and online activity. Darknet markets were centralized. Here, merchants and buyers could use pseudonyms to exchange products and payments with cryptocurrencies. While identities remained private, very real reputations could still be tracked and escrow platforms could be established. There was still a reliance on FedEx and UPS for shipment. The hierarchy of merchants was relatively flat, often giving single members the responsibility of closing the sale, which became exhausting. It was also hard for buyers lacking technical skills to join the overlay network. Eventually, law enforcement began to take down the sites and arrest participants offering these black-market services.

From 2017 onward, new types of online black markets emerged to solve problems faced in the "Silk Road." Merchants began operating on invite-only channels widely available through encrypted mobile messaging systems like Telegram. As a result of being widely distributed, merchants were less vulnerable to system takedowns. Merchants could also operate by bots, streamlining sales and transactional communication processes, without any direct human involvement. Meanwhile, consumers had much better access to the system. Delivery of goods could now be facilitated with dead drops, where items are left at a single public location for the customer to pick up. No personally identifiable information needed to be exchanged, but reputation can still be managed through pseudonyms on the chat forums. As the messaging channels facilitated more structure than darknet forums, they also opened up the opportunity to create hierarchical systems, in which merchants could employ a procurement layer, sales layer, and distribution layer in order to run their business more efficiently.[27]

The new model began to scale outside the black market rapidly. Decentralized servers, encrypted transactions, and retained control over one's own data began to entice those looking to post content without fear of censorship by the government or concern over their data being sold for advertising dollars. Others simply used the benefits

---

27 *Dropgangs, or the future of darknet markets.* (2018, December 26). Retrieved from https://opaque.link/post/dropgang/

of anonymity to chat privately, while a whole new force of merchants began to sell a wide range of goods and services that didn't require transaction fees.

The power of encrypted, decentralized, digital systems should not be lost on us. These networks are allowing consumers to better fulfill one another's needs. In doing so, they are inspiring new forms of decentralized labor, ones which allows individuals to operate freely. It is only a matter of time the data passing through these networks will be used to forecast and regulate value within micro-communities. The importance of this system is that it sets the foundation of Unitiism by establishing a single mechanism to:

- Forecast demand in a region.
- Produce goods and services for individuals by individuals.
- Allow individuals to benefit directly from their contributions.
- Obtain market feedback to understand the value of a good or service.
- Coordinate the delivery of a service.
- Allow members that create better goods and services to be rewarded with more opportunities to consume goods and services.
- Calculate the value of goods and services through a central technology-based, unbiased, body rather than a dependance on economic planning by any party outside of the community.
- Ensure equitable labor.

It is that freedom — the same freedom that many seek from the governments and enterprises they rebel against — that drives a projected 64 million people to work in the Gig Economy: 28% in professional fields (e.g., accountants, legal consultants) and 26% in creative jobs (e.g., writers and graphic designers). In the Fourth World, that sense of freedom will couple with the desire to serve the community, and it will orient around newly created cities.

It is worth understanding where, why, and how these new cities will form because, within them, we will be positioned to focus on Fourth World objectives.

Cultivating the right materials for machines

The natural resources we cultivate from the Earth today can be the ones we physically integrate with as cyborgs. It is the Earth's resources that can initially be used to encase Galactic and Aquatic Uploaders in hostile environments that are uninhabitable today. It is also the Earth's resources that can consume the consciousness of those that branch off to become Mind Uploaders. And it is the Earth's resources that can protect them against atmospheric changes.

Organizations, that we can simply call **Fourth World Entities**, will organize around natural resources to cultivate, preserve, and transform raw materials into components needed to reshape the three new fractions of humanity.

It should be noted that Fourth World Entities will not own, manage, or organize any goods or services needed to sustain a community. Unlike other institutions that look at readdressing a primary or secondary consumer need as the end goal, Fourth World Entities will look beyond the needs of today to design the blueprints for advancing Fourth World technologies. In doing so, they will remove competitive pressures and individual motivations that drive the demise of governments and existing enterprises today. Communities will form around them. It will work as follows:

1. **The public will support Fourth World use of natural resources**

    Fourth World Entities will be trusted to set up new facilities in areas inherently rich with natural resources. Although these areas are most desirable by governments and capitalist enterprises,

Fourth World Entities will be given priority to cultivate them.²⁸

We can look at why a Fourth World Entity would be chosen to secure hafnium, as an example. Hafnium is used in both processors and control rods for nuclear reactors. It's properties absorb neutrons, and

---

28 Disk storage has the power to store all that occurs for humanity. Disks are built using metals such as copper, aluminum, magnesium, and zinc as well as metalloids such as silicon. Alloys are also a key component in hard disks. The highest-performing hard disks use much rarer metals, though, such as ruthenium and neodymium. Data centers, the places that store large amounts of data, are highly inefficient though. They heat up quickly and so they need a lot of electricity for cooling, which pulls resources from the Earth. But data is useless unless it can be stored to be used bycomputersSoftware code gets translated into a combination of 0s and 1s, which represent low and high voltages. When voltages are applied to materials, they can change their physical properties. The process is done through microcontrollers called firmware, and data is transmitted through electronic circuits usually made from silicon. Plastics, which are created using coal, natural gas, and crude oil, are capable of creating complete electronic circuits. And then there are circuit boards, the things that connect the components. These are made of fiberglass, which is also produced from silicon. Next are the machine parts that can integrate with humans to complete the communication loop between the two bodies. They would use newer technologies that require more advanced materials. Artificial sheet material that modulates the behaviors of electromagnetic waves through plasmonic metasurfaces with sub-wavelength thickness are emerging. We see nanoscale devices such as graphene transistors, stretchable electronics, fiber and fabric batteries, and other sophisticated and artificially produced materials that don't face the same limitations as natural materials. They possess superior properties, such as flexibility, adaptability to environmental changes around them, portability, sensitivity, wearability, implantability, and stability in gases and water not found in today's most common materials. However, they also require a ton of energy to produce. The availability of these materials can be affected by government regulations that are formed as a response to environmental changes. For example, the environmental effects of mines, including loss of biodiversity, erosion, and contamination of surface water and soil, can be significant enough to result in governments shutting them down. Yet, many metals involved in technology, and rare-Earth elements, are simply byproducts of the production of more common elements like aluminum, copper, lead, and zinc. So, if these processes are reduced or stopped for any reason, then the source is cut off. Therefore, mining these materials important for the preservation and further development of technology.

show good mechanical and corrosion resistance. It's melting point is 2230°C and boiling point is 4600°C. It can survive atmospheric changes and can be repurposed to advance the Fourth World.[29] If governments publicly acquire large amounts of hafnium, however, they can send signals to the global community that control rods may be in development, which in turn raises tensions between nations and bodes poorly on government officials looking for reelection. Voters rarely favor political conflict. Meanwhile, providing these resources to Fourth World Entities for the purpose of protecting humanity can help realections.

2. **New entities will form, outside large cities, attracting new talent**

In 2017, Facebook announced plans to invest $750 million to build a data center in New Albany, Ohio, to store photos, videos, and additional digital content from its two billion users. This was after Amazon launched three cloud-computing data center sites in Ohio as well.[30] Both giants chose Ohio for its low risk of Earthquakes, hurricanes, and tornadoes. Ohio also has a duct system for broadband service and a robust high-speed fiber-optic network. Building the new centers creates new construction jobs for the community.[31]

---

29 Other more common materials may not run out, but recycling them from old electronics may become easier. The Ocean Cleanup Project is already beginning to gather, sort, and sell garbage that has accumulated in great patches in the oceans. Most of this garbage is non-biodegradable material such as plastic and metals, which can be repurposed for further technological endeavors.

30 *Facebook Officially Unveils Plans For $750 Million Data Center In New Albany.* (2017, August 15). Retrieved from http://radio.wosu.org

31 Duct systems are conduits or passages used in heating, ventilation, and air conditioning to deliver and remove air.

In addition to location and infrastructure considerations, most data centers are built with fire suppression systems, flood control, and Earthquake protection systems. They are forced to adhere to compliance standards on security, availability, and processing integrity. Managing these conditions requires different skill sets, creates even more jobs, and often brings in talent from all over the world. In fact, these jobs bring together some of the sharpest and most motivated individuals.

Fourth World Entities will begin to organize around locations that are conducive to the survival of machines. They will act similar to the enterprises that are currently building data and technology centers in locations with the safest environmental conditions. They will drive new talent to work on their endeavors.

Communities will slowly form around Fourth World Entities, allowing us to reinvent our territorial design. We will be able to remove geographic impediments that caused communism and socialism to fail in the past. These communities will be built with public good in mind. New communities will provide an opportunity to engineer locations with shared food banks, shared housing, shared everything. It is worth examining their design in depth.

New cities and transportation

There will be one exception for Fourth World Entities to own and manage goods within a community. Fourth World Entities will optimize communication, energy, and mobility technology used in systems today to advance digital, physical, and biological development for the future by building a roadmap that uses assets to their capacity. The products they create can be used within the new cities for testing and exploration with community members.

Fourth World Entities may set up new cities with autonomous vehicles that serve consumers today. Autonomous vehicles will be electric and can be built to optimize energy through nanogenerators. Nanogenerators can create energy-generating car wheels. This technology can also be extended to fuel the systems that will power exploratory vessels for Galactic Uploaders, once they land on new planets. Similarly, technology that powers vision systems for autonomous vehicles, today, can be transformed to scope new environments in the future — the same vision systems can be embedded in Distinct Intelligent Machines to sense new planets.[32] We can also see how the supercomputers that power autonomous vehicles can help with decisioning capabilities for Aquatic Uploaders — they can quickly search a database of ocean life and make decisions in real time.

New cities may be designed with looser regulatory conditions, and consumers will ride these vehicles for free in exchange for taking on the risk that serves as Fourth World Entity learnings. Citizens will participate in other Fourth World research projects as well. This may mean that they will integrate with new sensors to read their biometrics. They may be tracked using city-wide cameras with facial recognition. As Fourth World Entities use cities as test-beds for their research, they will both learn and enhance the life of the citizens and they will enhance the new technology.

The formation of new cities, using new technologies, promises deep customization and personalization for each individual. Each city will be different, but all will be desirable, helping like-minded individuals come together to form tightly knit, almost tribal, communities. Not all cities will have autonomous vehicles.

Cities will be small, and those that aren't born with autonomous vehicles will have tunnels and pathways connecting every building. Elevators, gliding walkways, hoverboards, and other micro-transportation modes will flood the neighborhoods.

---

[32] Systems such as LiDAR and RADAR which use light waves and radio waves, respectively, to detect the distance and shape of an object can be used in the Fourth World to create renderings for machines to help inform humans of physical conditions as they explore new environments.

Getting to and from these new cities will become easier and easier as transportation and mobility changes globally. Autonomous vehicles will also certainly exist outside of Fourth World cities. They will expand urban sprawls in established markets. Many have predicted that thanks to the sheer joy of a commute in an autonomous vehicle, people in established markets will move further away from urban cores, making it even more possible to join cities that align with their ethos.[33] Over time, autonomous vehicles will either exist for long range travel or they will be enhanced for Uploader testing.

Coupled with macro-sustainable living trends that further enable micro-communities to be more self-sufficient, these city trends are largely supported by changes in the logistics business. The logistics business has already started to feel a shift.[34] A new hub-and-spoke model will emerge as owner-drivers and transitional commuting passengers begin solving last-mile problems on their own.

While in transitional phases, major suppliers of goods will still dominate in a monopolistic fashion, just as they do today. It is estimated that 25% of all e-commerce business in the USA is still conducted by only five major online retailers: Amazon, Walmart, Apple, Staples, and Macy's. E-commerce will support the demand within new

---

33 Headlines declare that two-thirds of the population will live in the urban core by 2050. The trend is primarily expected in emerging markets such as China, India, and Nigeria. There is a threat that the indigenous, ethnic minorities, within these countries, looking to major cities for opportunities will jeopardize their heritage and forego chances to spread knowledge of their history. Yet it is impossible to ignore that, just as centralized digital marketplaces are losing ground to distributed systems, we are at the brink of a global reshuffle, breaking ourselves up into self-defined communities.

34 As of 2016, Deloitte reported that the global trading volume was falling at a compound annual growth rate of 2.5% over the past five years, with a sharper decline of 12% between 2014 and 2015. They predicted that, if this trend continues, negative pressure would be placed on the demand of transport services. In addition, pressures from rising costs and higher consumer expectations for quick delivery will drive fundamental change in logistics business models and operations. Smaller logistics platforms will not be able to keep up. There will be consolidation of businesses. The larger fleets will have heavy commercial vehicles, best used for long-haul commutes and not particularly suitable for delivering consumer goods to remote areas.

communities as they transition from being globally reliant to self-sustaining. Truck drivers will still be required in the transition. However, as automation of transportation moves jobs away from people to machines, it pushes workers who once manually labored in this field (i.e., drivers) to either retire or to gain new skills. Autonomous trucks will operate without embedded drivers on long-range hauls. They will be monitored by drivers from a base-station inside the new cities. And, drivers will only enter the trucks once they reached the city limits to deliver the last mile.

We can imagine the mashup of Uploader-testing and automation in new cities. New skills will be easier to learn as automation replaces more and more jobs. Cardiologist, Geneticist, and Digital Medicine Researcher, Eric Topol, discusses how technology can help doctors better scan medical images in order to diagnose irregularities, such as cancer, in our bodies. He hypothesizes that, as a result of this technology, doctors will be needed more for their bedside manner — for their ability to care and nurture patients — rather than for their ability to review lab images, as that part of healthcare can be achieved by machines. Taking it a step further, we can imagine that doctors may eventually become doctors because they choose to integrate, as cyborgs, with devices that help scan medical images better. We can imagine that, while image processing will be automated, the human mind will still be needed to pass judgment — the doctor's eyes could be replaced with a scanner to detect MRI scans and diagnose cancer with 100% accuracy. What automation holds is the ability for us to dedicate ourselves to more empathy — to connect with one another on a deeper level.[35]

Signs of new cities are already developing to set this foundation. Bill Gates recently invested $80 million in building a new smart city just outside Phoenix, Arizona. The city of Belmont promises eighty

---

35 Of course, we won't completely understand how machines diagnose the images. Topol explains that machines can recognize a female v.s. male by looking at their retinas, but human eye doctors can't. This is because we can't deconstruct the deep neural networks that machines use to learn the differences, but we will learn more about this in later chapters.

thousand residential homes supported by retail, commercial, and industrial space, along with 470 acres reserved for schools, and another 3,800 acres of open space. The city will, of course, plan to integrate autonomous vehicles and high-speed digital networks into the infrastructure.

Gates' investment pales in comparison to that of Abdullah bin Abdulaziz Al Saud, former King of Saudi Arabia, who has invested $80 billion in a city just outside the Islamic Mecca on the Red Sea. The King Abdullah Economic City plans on housing two million people, of which 40% will be under the age of fifteen.[36] It will have one of the largest seaports in the world.

Within new cities, members can be awarded stipends. This can work similar to the concept of **Universal Basic Income (UBI)** pilots being instituted around the world today, where $1,000 a month is awarded to low-income members of a community in replacement of other government-subsidized programs such as welfare. Yet, in Fourth World cities, everyone will receive the same stipend for the purpose of launching and maintaining a good or service that can be shared within the community.

The stipend will be managed by a digital system, initially created by one Fourth World entity, but used by every city. It will be managed by a single member who does not own the system. Rather, the member is responsible for operating and maintaining the system. Every stipend will be recorded on a global, immutable technology (i.e., Blockchain). It will ensure that there is no poverty within the community by giving each member a chance to serve others. We will talk about how the stipend will be funded in the next chapter. For now, it is important to know that the notion of stipends will only exist as a mechanism to transfer money from the world we live in today into the Fourth World. Eventually, we will operate without stipends. The Fourth World will survive on true value.

---

36 *Saudi Arabia Faces Population Pressures.* (2003, May). Retrieved from https://www.prb.org

True value

In the Fourth World, individuals will operate through the exchange of a credit system, and it will not lend itself to taxation. All transactions will be managed and recorded by immutable technology that measures the value of the goods or services based on the exchange of credits that other individuals within the community are willing to make. Meaning, technology will govern the value of assets produced by each member and dynamically adjust based on each member's bartering power. Each city will constantly be calibrating the value of any good based on its own value system. Each community will create its own notion of true value.

For example, if one community member produced corn and another member produced beer, when the beer maker would initiate a transaction to request twenty stocks of corn for a keg of beer, the immutable technology would value the twenty stocks of corn and one keg of beer at equal value. If the beer maker then traded a keg of beer for a pound of clothes, the technology would then calibrate the pound of clothes and corn as equal. If the producer of the clothes then traded a pound of clothes back with the corn maker for five stocks of corn, the system would calibrate the stocks of corn at higher value. Credits would represent the current value of each individual unit of corn, similar to a dynamic unit of currency, and the value of the credit could be used as bargaining power in future transactions.

To best understand how a system can survive on credits alone, first one needs to know the history of currency. Early on, all currency was equal. Then it wasn't. And now, we are creating all new forms of it. Monetary history shows that politics, rather than economics, has been the chief driver of currency decisions. This usually happened to meet the demands of expanding international trade conditions. A single currency followed the expansion of its attendant political power (e.g., Roman, Chinese, and British Empires introduced new coins as they conquered and joined regions). When politics were stable, businesses benefited from the single currency; however, when they were unstable, alternative currencies were introduced. With larger communities, bartering was annulled as the problem of "the double coincidence of wants" arose. If people didn't want what was available, it was impossible to carry out an economic trade. Currency provided a medium that everyone was willing to trade for. It had high intrinsic and storable value. It required a common social understanding, a single government, and a wide trading area that could benefit from its value.

As instability within empires separated and isolated communities, these commonalities became less common. Everyone started making their own currency coins. This created its own set of problems, as struggling governments started to debase currencies to get more tax dollars from citizens. Not only did this create uncertainty and make citizens poorer, but it also inhibited trade. There had to be an exchange and an exchange rate, which were both good because they set up some level of control. However, these also required the integration of governments.

More countries and currencies exist today than ever before. While the international economy has become more integrated than ever, the political integration that was once required for financial integration is no longer necessary.[37] New communities are forming on the basis of alignment of value, free of government borders, as governments change their role in society. New currencies are forming within these communities. There is no standard metric by which to measure them.

An example of this can be seen with the rise of digital capital. Platforms like Steemit simply give away their proprietary cryptocurrency to users generating blog content. Apps are giving away electronics in exchange for fitness and health-tracking data. Then there is face-value capital. Instagram Influencers get free vacations in exchange for posting praise for an island resort. In Silicon Valley, housing is given in exchange for labor. There is even energy capital. Some utility companies exchange energy with electric bike companies who establish grids to transfer energy back and forth between charging stations (i.e., energy from unused batteries is sold to major utility companies who sell the same energy back when it's needed).

---

37 Taylor, B. n.d. *A history of Universal Currencies*. Retrieved from https://www.global-financialdata.com

This currency is non-taxable by government bodies (unless, of course, an individual transacts out of the system, but this will not be necessary as the Fourth World evolves).

What credits in the Fourth World hold is true value. Credits do not go up and down with market conditions. They depend on deeply fulfilling one another's wants and provide access to all members within new cities based on what they produce. They allow for a new renaissance that restructures our relationship to life.

In new cities, *individuals can serve one another* within communities, absent of income pressures, as our *global investments shift focus* to support the Fourth World.

CHAPTER 6: **UPLOADING HUMANITY TAKES** *BIG BETS*

"I have no intention of making small bets."
— *Softbank CEO, Masayoshi Son*

Egyptian Pyramids

The Egyptian Pyramids were built right before the Egyptian decline. And that's what society does. It accrues it's greatest wealth right before it dies. We can use our wealth to build a massive monument in which our bodies will rest, or we can do something different. We can reinvest our money on a few big bets that save humanity.

We could only have about 80 years before a mass extinction event. And, while the creation of the Fourth World will take well over 80 years to develop, we can't afford to fail by not investing in it now.

We have to ensure that any Fourth World Entity is a Big Bet Entity. Big bets will be expensive. To realize these bets it will take participation multiple sources.

A new trend in funding social good, called **Venture Philanthropy**, is already forming. Venture Philanthropists are a form of **Venture Capitalists (VC)** that use investment dollars, aggregated from many sources, to achieve a high-engagement and long-term approach to creating social impact. For Fourth World Entities, it requires that the social impact moves toward Uploading.

Funding the Fourth World

Without funding a vision, results are not manifested.

In the late 1950s, human exploration of the Moon began with a race between the United States and the Soviet Union. The results of their efforts demonstrate what would happen if we fund technological advancements rather than the The Fourth World vision in its entirety. We can compare the results of this funding with the funding of Jeff Bezos's vision for intergalactic travel to understand why only a few significant investments can set the course for our evolution.

The United States spent about $25 billion to send a man to the Moon. This equates to about $180 billion in today's money. It is more money than the USA spent on any single effort, with the exception of the Vietnam war. Meanwhile, the Soviet Union spent less than $10 billion to send their own aircraft to the Moon. Both nations had tons of failed attempts, but neither gave up.[38] It was a race that defined the technological strength of nations at odds amidst the Cold War. Eventually, the USA won the race in 1969. But that was over fifty years ago, and what has really come from the mission? The space station grounds that launched the US ship to the Moon from the Kennedy Space Center in Florida remain unprotected from climate change and are quite literally washed away.[39] Moreover, while scientists learned that conditions on the Moon would make it difficult for humans to inhabit, the US has not pushed through impediments. They did not set up base stations on the Moon, which could be used to either continue exploration from the Moon or, perhaps, make inhabiting the Moon possible today.

---

38 *Race to the Moon*. Retrieved from http://www.historyshotsinfoart.com

39 Lepore, J. (2019, June 14). *Fifty Years Ago We Landed on the Moon. Why Should We Care Now?* Retrieved from https://www.nytimes.com/

In contrast, Jeff Bezos, who created Blue Origin to chart a course for a massive galactic journey, is funding an effort absent of competitive drivers. He is doing so in order to "return to our natural state with the limitless energy of the solar system."[40]

As part of Bezos's vision, we will reinvent our way of living. On Mars, we will each weigh 38% of what we do on Earth. Our hearts will be lighter. We will need less, forgetting our essence as human beings defined by our mindless habitats. Shipped off to an uncolonized land with the tools necessary for survival, we will have no established economy to overcome. There, we could build a land absent of judgment and racial stigmas. All inhabitants would be one people, defined by a more profound commonality. There will be endless things to discover and challenges to overcome. There will even be a euphoric sense of love stemming from daily adrenaline.

What we can see by comparing the motivations of governments and Bezos is that investing in a big goal, without a vision, can lead to wasted efforts. Following through on space travel must be driven by an effort that doesn't stop until humanity is reset. And, only members of the Fourth World will be positioned to culminate this effort. We must bet on Fourth World initiatives.

Like any big bets with a vision, Fourth World innovations will be exorbitantly expensive in early stages. Funding them will require pooling resources from many groups. Venture Philanthropists can help with this. Here are just a few other sources they can tap into for garnering funds to make this possible:

1. **Existing government programs.**

    Unitism will require less support from governments to fund secondary consumer needs. While many governments are in operational debt, changing economic conditions will leave many bodies with a surplus to relocate to Venture Philanthropist funds. To make this transfer possible, funding of programs

---

[40] Heffernan, V. (2019, January 19). *Love and Rockets.* Wired Magazine

such as the ones that follow, will need to cease (these examples are in the US, but the notion can be applied globally):

- $1.04 billion from the Department of Transportation for San Diego to expand the city's trolley service by 10.9 miles to serve 24,600 people might ride the trolley each day.[41]
- $1.7 billion to maintain 770,000 unused schools, firehouses, offices.
- Any animal testing program including:
    - $3 million from the National Science Foundation to observe shrimp running on a treadmill.
    - $3 million from the National Institutes of Health to inject hamsters with steroids and watch them fight to see if the drugged rodents were more aggressive when protecting their territory.
    - $518,000 in federal grants to study how cocaine affects the sexual behavior of Japanese quails.[42]

1. **Private endowments.**

There are also private endowments held by thought-leading bodies such as Universities that can contribute towards funding Fourth World Development. For example, Harvard University currently has $37.1 billion under management. Although this is not enough to fund the Fourth World, it can contribute to Venture Philanthropist funds focused on Uploading. And, as members will

---

41 *5 Outrageous Ways the Federal Government Has Wasted Your Money (Part IV)*. (2018, July 6). Retrieved from https://americansforprosperity.org

42 Laliberte, M. *11 Bizarre Things the U.S. Government Actually Spent Money On*. Retrieved from https://www.rd.com

no longer need things like pension as they move into new cities, excess funds will become available.

2. Behemoth corporations.

Between the $961.3 billion that Apple made, the $281.3 billion that Nestlé made, the $238.1 billion that Walt Disney made, or even the mere $153 billion L'Oréal made, corporations are positioned to fund Fourth World innovations.

However, no single source can fund the Fourth World on its own. All in all, many parties must come together.

So, why would governments, endowments, and corporations come together with Venture Philanthropists to completely deplete funds from today's world into Fourth World Entities? Because if they don't deplete their funds in this direction, their funds will be depleted, naturally, in another.

A few entities *will* create a mechanism to shift consumer spending from individual consumption to a credit-based system that uses technology to manage true value. A few organizations *will* believe in their long-term strategies. They *will* invest in their future (sometimes for short term personal gains, loosened long-term investment returns, or inaccurate projection models).[43] And that's exactly what Lyft and Uber have positioned our society to do.

Lyft & Uber IPOs

When Lyft and Uber submitted the documents to file their Initial Public Offerings (IPOs), allowing their initial VC investors to liquidate, they

---

43 In economics, a model is a theoretical construct representing a set of variables and a set of logical and/or quantitative relationships between them. It can help investors guide their investment decisions based on things like the profitability of a company, how weather conditions impact crops, or how government relationships impact tariffs and their resulting prices on goods. Yet, the further out these variables our, the less accurate a model becomes.

clearly stated that they might never be profitable. Public transportation isn't profitable either. Yet, it has historically been provided by governments to fulfill a subordinate consumer need. The difference is, as Uber and Lyft IPO'ed, they gave consumers the ability to judge services and the individuals that provide them. Drivers aimed to deliver more value to riders so that they got higher scores and were offered more trips. Riders began enjoying the services and wanting to share more rather than consume (changing the notion of consumer acquisition, for acquisition sake). This is a stark contrast to using tax dollars through government bodies because, there, consumers have little say in the value of the services or people that provide them. Moreover, Uber and Lyft began providing a system that encouraged more use by giving away free rides (i.e., credits).

Why did the investment community fund a non-profitable business, and why do they continue to support the stock? Well, an unprofitable company doesn't mean investors can't profit. Those that invested in Uber and Lyft early on saw returns on their investments when they were worth more in the IPO than the original purchase price. Once the public offering hit the market, other investors subscribed to a "winner-takes-all" model: the idea that if one company achieves critical mass, it will put all other companies out of business. And so, they hold onto their investments for the long haul.[44] Then, there are those investors that make money by playing the market through turbulent times, as they buy low and sell high at momentary spikes and valleys. Platforms that use technologies to allow one individual to serve another, absent of the platform's profitability, are shifting the flow of money from a global body of investors to support communities. And, investments can continue to shift to fund Fourth World Entities.

What do Uber & Lyft IPOs do for governments? Well, as a result of institutional and public investments, the platforms began lowering the prices of rides. This resulted in Uber and Lyft becoming favorable

---

[44] Although this is not the case with Uber and Lyft today, in some cases, winner-takes-all strategies cause come companies to receive more funding than needed. The investment dollars exceed the market cap. Meaning, they can actually get more money from investors than they can earn from the market today.

competitors to traditional models for public transit. This could easily shut down local bus and train systems, as they lose their reason for existing. So, governments have already begun to address the growing losses that are starting to result from operating the transportation system with little revenue from ridership. Allocating more tax dollars to a program with declining ridership is not an option. Taxes also won't go away in the short term. So, governments have begun to repurpose tax dollars by subsidizing Uber and Lyft rides for low-income members of the community. Here is a second example of a shift in funding that can support both Unitism and Fourth World Entities.

However, the role of institutional investors and governments, in funding the gig economy, is very different than funding Unitism. It is simply an entry point for supporting communities, who support one another. Over time Unitism will not require monetary transactions because individuals will exchange credits with one another. The case with Uber and Lyft simply sets a precedent for funding the technological advancements needed to make big, dramatic changes in our socioeconomic structure.

Funding will mostly go to new entities, with a strict focus on the Fourth World initiatives that fork humanity off into the three paths of Uploading. Fourth World Entities will establish contained cities that welcome a community of research-subjects (i.e., suitable testers for Upload technology) who will migrate there to receive true value. Unitism will not require monetary payments between individuals.

So, how can companies like Lyft and Uber pivot their situation to become Fourth World Entities? They can garner even more investment dollars from VCs. This is shrewd financialization. Uber can use VC funds to spin-off of Uber Copter, an air service, that can be used to build a blueprint for advancing Galactic Uploading. Meanwhile, Lyft can spin-off it's autonomous vehicle division to advance other Fourth World initiatives.

Uber and Lyft only represent a sample of companies that are positioned for spin-offs. There are, of course, many large enterprises with excess funds today, and they too can spin of Fourth World Entities. In fact, their excess funding implies global societal confidence, which can be leveraged for new endeavors.

Gateway Members

In addition to Fourth World Entities and the community members that surround them, there is another, very important, player that will contribute to the origins of the Fourth World. These could be individual beneficiaries of IPOs who are offered paper shares when they join, or start a company, as part of their compensation package.[45] However, they could also be part of the 2,153 living billionaires that are cumulatively worth $8.7 trillion.[46]

These individuals will move on to new cities where they will work with Fourth World Entities on building a blueprint to transition the current general-purpose platform into one that Uploads humanity. There, they will become patrons of the community, forming projects that support infrastructure. They will create services that help community members live a more fulfilled life. And, they will contribute to the city's version of a Universal Basic Income pool of funds, by equally distributing their personal earnings, from a prior life, to all members of the community. These individuals will be called **Gateway Members**.

Gateway Members will be interested in providing economic equality. This will be their motivation for converting financial assets into communities, to provide things such as free housing or sharable

---

45 IPOs provide opportunities for individuals who were promised stock in startups to "cash out" on the exchange of private to public ownership. Their paper promise is made liquid. When most companies IPO, they split the value of their existing stock to issue additional stock. This dilutes the value of each stock to its owner. Some enterprises IPO using a tactic called Direct IPO, otherwise known as "DIPO," which does not split the stock. Along with investors, this can also make early members of the company wealthy. SoftBank funded enterprises are known to do this, and many see SoftBank, a Japanese company, doing this as a way to extract cash from international markets.

46 (2019, March 5). *The Richest People In The World*. Retrieved from https://www.forbes.com

scooters. Here, governments have an opportunity to incentivize these efforts by providing tax breaks for individuals who move profits from IPOs into Fourth World Universal Basic Income pools.

Like fiefdoms, Gateways will maintain allegiance to select Fourth World Entities. They will not churn from company to company, as they may have done in their former life. This is because, unlike the enterprises that employ individuals today, Fourth World Entities will be viewed as organizations with a meaningful, long-term purpose. Gateway Members who join them will see it their life's work to deliver on the aligned purpose, absent of any personal gain.

And, like fiefdoms, Gateway Members will act as liaisons between the larger Fourth World Entities and community members. They will coordinate between communities to procure goods and services unavailable to their community. They will be trusted. They will know the members well. And, they will work with Fourth World Entities to make recommendations on who should be selected, within the community, for early Fourth World research and testing.

Ethics

We can see how opportunities exist to fund the Fourth World by bringing together the likes of governments, endowments, and corporations, with to Venture Philanthropists. We can see how institutional investors will play a role in transitioning funds to big bet entities as well. And, we can imagine a world where wealthy individuals augment Fourth World developments. The questions then become, what should be funded, and how do we know that we are on the right track?

Big bets will be oriented on:

- Cyborg development to advance our material form
- Biomedical engineering of our current material form
- Dome creation to support Uploaders
- Advanced communication channels between space, land, and underwater

- Life support technology such as food sources

Since the Fourth World won't be built on profitable business cases, we will need a mechanism to validate these bets; therefore, showing constant proof-points that these bets can, in fact, advance humanity will be necessary. As mentioned before, communities that form around big bets will serve as test-beds. And, for testing to be successful, we will have to take some risks.

Fourth World Entities will be founded on **Kantianism**, a moral theory in which the rightness or wrongness of actions does not depend on their consequences, but on whether they fulfill our collective duty.[47] Deontological, duty-based ethics are concerned with what people do, not with the consequences of their actions. So, under this form of ethics, one can't justify an action by showing that it produces good consequences; the ends do not justify the means.

Fourth World Entities will have a duty to fulfill the Fourth World mission. They will be set up to create the ability for humans to transcend space and time. While our relationship to life will be centered on both individual enjoyment of it and the preservation of it for all mankind, entities that guide our new renaissance will have to make certain decisions in pursuit of their duty. This means the associated actions of those entities may come at a cost.

While performing the duty to preserve ourselves and enjoy life, entities may provoke unsavory consequences, such as the death of animals or harm to people. Duty-based ethics argues that humans should do the right things, even if that produces more harm (or less good), than doing the wrong things.[48] This can be an understandably uncomfortable theory for many.

However, as mentioned above, the Fourth World will experience a transition period where capital funding is still required to create the new technology for Uploading. For entities to procure this funding,

---

[47] *What is the Fourth Industrial Revolution?* (2016, July 18). Retrieved from https://www.youtube.com/watch?v=kpW9JcWxKq0

[48] *Duty-based ethics.* Retrieved from http://www.bbc.co.uk

they will have to show signs of success to funding parties in the form of research advancements, while enticing community members to participate in experimentation. For example, depending on the goals for Fourth World Entities, members of one community may be invited to integrate with laser eyes. Members of another community could be fit with exoskeleton shells. The success of the tests will serve as duty-based proof points to investors.

Society and Fourth World Entities will have to accept that accidents will occur as part of the experimentation. Yet, with each mistake comes new possibilities to take corrective actions that will advance our learning and technology. While Mind Uploading is being developed, there may be issues with the brain mapping process. Or, as devices are implanted into the human brain, extracting memories could impact an individual's experience with life. However, the mere success of the extraction could prove that a Fourth World entity can, in fact, advance the technology to fulfill its duty.

There will be an exchange between Fourth World Entities and their communities. This shouldn't be forgotten. For example, while individuals in one community may be required to submit genetic samples, they may benefit from acting more selfishly with their use of environmental resources, as Fourth World Entities will act on their behalf to save humanity. With this model, a utopia can emerge to serve the underpinnings of our society because the exchange of value doesn't depend on monetary exchange alone. It will provide *more* opportunities for *more* individuals to actively participate in society, by providing *more* value to one another. Enjoyment will come in magnitudes, rather than increments, as communities can rely on big bets to advance the transition towards the Fourth World.

CHAPTER 7: UPLOADING HUMANITY TAKES *WAR TACTICS*

"In wartime, truth is so precious that she should always be attended by a bodyguard of lies."

— Winston Churchill

A silent war

We are fighting a silent war and nobody even knows it. The war is competition. Competing ideals. Competing goals. Competing individuals. All forms of competition work against the future of humanity.

It is arguable that competition drives innovation. It does drive innovation in a capitalist world. But, capitalism won't save humanity.

It is vyingly arguable that innovation can exist if society is presented with a challenge to redesign the material form of humanity. And competition does nothing for this challenge.

There is only one way to fight this silent war of competition and win, and that is to build giants.

A lesson from Mr. Washburn

The story of General Mills shows how to build giants and work together to serve the objectives of a new world direction, particularly with material preservation.

It was in the late 1870s in Minneapolis, Minnesota. Builders of the great American nation — lumbermen immigrants — were eating rancid wheat. Begging it to produce the fuel they needed to perform their manual labor, they were blinded to the fact that the most nutritious elements of the wheat were lost in the milling production process.[49] Such was the case until Cadwallader C. Washburn founded General Mills.

Washburn was from modest beginnings. He created the revolutionary technology of gradual reduction processing. The technique used steel rollers along with purifiers and an exhaust system to process grains. It was not only able to produce nutritional premium-quality pure white flour, but it did so rapidly.

But, with success, competition followed.

Charles A. Pillsbury, also stemming from modest beginnings, created the Pillsbury Company across the river in Minneapolis. He started his company in the 1870s, almost twenty years after Washburn. Pillsbury hired Washburn's employees, at higher wages, to immediately learn and use the new methods pioneered by Washburn. Through this use of employee poaching, Pillsbury's A-mill soon became the largest in the world, breaking production records.[50]

Pillsbury Company wasn't Washburn's only competition. A secret war was being fought all around him. Minneapolis primed itself as open to rivalship, as it had a long standing nemesis with the neighboring city of St-Paul. The rivalry was so bad that, at one point, architects from one city were banned from working in the other, while people

---

49 Watts, A. (2000). *The Technology that Launched a City*. Retrieved from http://collections.mnhs.org/MNHistoryMagazine/

50 Kend, Aispuro, and Liang. (2003). *Engines for Growth: Pillsbury Brings Potential*. Retrieved from http://economics-files.pomona.edu

kidnapped and arrested census takers of their opponent city to keep it from outgrowing itself.

In 1878, things got worse for Washborn. Disaster struck when General Mill's facilities suffered a physical explosion. But Washburn had a duty. Vowing to rebuild his plant and protect his assets, he adopted new technology, including ventilation, to make mills safer. He hired the best engineer: a blind, Bavarian who spoke no English. Eventually, he got the job done.

Then, Washburn devised a plan to build a giant.

First, Washburn eagerly shared the safety measures with all his competitors, including Pillsbury. Then, he set out to take more control over steel and electrical power, the very components needed to make railroad tracks — a key factor missing from the supply chain delivery. Washburn petitioned Pillsbury to take advantage of their joint position in steel to help him advance the construction of a railroad that could export both parties' products.

Washburn's plan eventually began contributing to the massive export of Minneapolis flour globally, putting Washburn and Pillsbury on equal playing fields, far above any competition. The construction of the railroad also bridged relations with St.Paul and Minneapolis, as St.Paul was eager to enjoy their former rival's success to help trade.

Washburn's then used his profits to grow his business through partnerships. He began acquiring smaller competitors.

Washburn died ten years later, but his successor continued to follow through with the final stages of his plan. They merged with several other regional mills. Eventually, General Mills became the largest mill company in the world. In October 2002, with the support of the government, they bought Pillsbury, putting the General Mills giant in a dominant market position.

What we can learn from Washburn, as he moved from production of flour to creation of a railroad system, is that creating giants kills competition. But, creating giants takes collaboration. Big-bet entities must all come together to align on dividing up the stakes in the future, as multiple new distinct products and services must be built for Fourth

World survival. Lastly, what we can see from Washburn, is that big bets can arise from anyone, anywhere. It should be noted that Big Bet Entities will not all originate from wealthy nations or individuals. This strategy must be formed on a global scale. Integration and partnership are essential, as are discussions on the lack of duplication and agreed-to ownership.

A new division of responsibility

Creative destruction, as discussed in the previous chapter, will make it possible to accomplish the initial transition from today's world to the Fourth. As creative destruction begins, we will need to ensure that we work together, as a society, towards common goals.

It will be important for us to limit duplication of effort, which is normally present in competition. Unlike the enterprises that exist today, competition will not be necessary in the Fourth World. Competing forces are only needed to offer incremental benefits to a consumer. However, incremental benefits to a consumer are not what the Fourth World is charged with. And so, there is no purpose for it. Big bets will be manifest in space stations on Mars and places that can't be fully reached by many people in our lifetime. Big bets will be achieved through seabases underwater and environments that may also not be fully inhabited in our lifetime. And, there are many other examples of the things we will need to build throughout the next few generations to make Uploading a reality.

Big bets will also require cross-integration. Multiple Big Bet Entities may come together to create things like genetics-altering biotechnology that connects directly to the Internet of machines. Dependencies will need to be mapped, and integrations will need to be tested.

Since these bets will require cultivating resources from this Earth, we must also be mindful of how we allocate resources across various Uploading initiatives. Materials will be harnessed from all over the globe. Yet, there will only be enough resources for a few big investments. Various parties will need to understand how to divvy resources most effectively to deliver on a common blueprint for humanity.

We can achieve this sort of collaboration if we divide up responsibilities. This managed division of responsibility could take form through a global and cross-disciplinary committee. This **Fourth World Committee** can envision and create a technical blueprint of the Fourth World. Fourth World Entities will be defined on the basis of this blueprint. Each will serve a different purpose, and each will be charged with creating distinct technology.

Entities involved in the blueprint should state a claim in the Fourth World and be required to officially share that claim with governments. This is not a vision statement; rather, it describes a responsibility that the company will uphold. The government can ensure its alignment with Fourth World goals and needs.

A few early enterprises can be selected by the committee to transform into Fourth World Entities, as part of a pilot program. To transfer funds from today's economy into the Fourth World, these selected parties can create an enterprise mitosis and engage in shrewd financialization. This will likely involve selected parties to divvy up territories that their current products and services support to capture a market fully. With full market adoption of current services, they will IPO to spin off new entities geared towards big bets on the Fourth World. These ventures will eventually be joined by publicly-traded goliaths, who will fund new long-term growth entities to leverage core capabilities in new ways, while visionary entrepreneurs, with track records of success, will receive exorbitant funds from Venture Capitalists to do the same.

The Fourth World Committee can also join forces with institutional investors to measure the success of pilot entities, and new entities, as they emerge. They will review research to validate that the big bets are working. As research is validated, they will find opportunities for collaboration across big bets.

Aligned with the claim, governments can help entities set up cities as testbeds. Entities can act to attract the best and brightest to fulfill the claim. By offering equity and benefits from an IPO that takes advantage of shrewd financialization to raise more VC funding for product development while spinning-off future bets, those they recruit to work in the entities will serve as Gateways who will reward the communities built around the city while keeping community members available for testing.

Lastly, we can begin to measure ourselves on the increasing number of needs-met for every human on the planet, rather than a select few. As individuals will assemble around big bets, we will need to ensure that value is distributed equally across each living member in society. All communities will distribute an equal portion to each individual. Individuals will join the community that aligns best with their values. In fact, greater distribution of wealth is both a signal of a successful roadmap and a true enabler for the path forward.

Regulatory matters

Fortunately, governments can help us put the war we are fighting to rest. They can force us to innovate as a joint society. This goes beyond transitioning tax dollars that were formerly used to serve secondary consumer wants into funds for society to advance innovation. They can form a partnership with the Fourth World Committee and create enabling conditions to reduce competition. They can provide regulations and interoperability standards for cross-collaborations. They can ensure that we take the necessary measures to transform humanity into a state of surviving any atmospheric condition. Here, governments, on a global scale, should patronize Fourth World Entities by regulating as follows:

1. **Regulate the profitability of every enterprise.**

    By enforcing the profitability of an enterprise, those that achieve high margins will be forced to build Big Bet Entities. We see how this can manifest itself by looking at Insurance companies that have highly regulated margins today. They usually charge higher premiums than they pay out claims. So, in order to keep margins according to guidelines, they invest in spinning off growth technology businesses. Meanwhile, those that can't maintain profitability guidelines will perish.

2. Divide up existing service territories between single entities (monopolies).

Governments that favor free markets currently subsidize enterprises by doing things like giving them tax breaks or providing ease with patent registration. However, intellectual property remains primarily within the country, and it creates an environment for enterprises to compete for self-sustaining businesses rather than focusing on outcomes that help humanity. These governments see this as a good thing and tend to create regulations that limit single players from dominating a market. They want competition to drive more benefits for consumers, but they should be taking the exact opposite approach if they want to advance the Fourth World.

Governments must remove anti-trust laws and any regulations that prevent single entities from dominating any part of today's consumer market. Some modern enterprises are already in a position to make big bets because they earn high revenue on a given offering. Yet as competition naturally exists a market, all enterprises must agree to divvy up ownership by territory or value chain fulfillment so that they achieve steady profitability to reinvest into new ventures that will not profitably advance the Fourth World. For example, Lyft could agree to service all of America, while Uber agrees to service all of the UK. Doing so will help both companies achieve a profitable state that will allow them to fund new, non-competing Fourth World objectives. Those companies that do not succeed in a given territory will cease to exist.

3. Maintain blueprints towards Uploading.

Governments must seek to understand all the components of the Fourth World. They should work with the Fourth World Committee to support the unique needs of Fourth World monopolies. This will require custom allowances for each entity. It will require also require establishing a support structure and an integration mapping with various service providers.

4. Incentivize new cities.

Although Gateway members will help subsidize Universal Basic Income programs, in the early stages, individuals may need additional incentivizes to migrate to the big bet communities. Governments should support a fund for each new city and provide an equal portion of the fund to each new member. This will seed Unitism with initial investments that can be used to help individuals create micro-offerings to other members within the community. These funds can arise from getting rid of existing government programs that support common welfare. Governments should also redesign economic development policies to encourage engagement within cities to support the validation of big bets by Fourth World Entities.

5. Ensure entities take advantage of today's general-purpose technology platform.

Governments should have access to an entity's intent and make available any communication channel, energy, or mobility asset to be leveraged for digital, physical, or biological advancement[51]. They

---

51 For example, one organization may be responsible for using silicon to create circuit processors while another may want to employ all the individuals with a specific data

should ensure that these intentions align with serving a Galactic, Aquatic, or Mind Uploading development cause.

6. Get adequate resources from any global territory.

This would require governments to allow entities to define materials, that are best suitable for their purposes, and allow mining of these materials at the potential cost of the environment and local economy.

7. Integrate services.

Governments should mandate that monopolies creating a Fourth World infrastructure integrate their services with one another and across the globe. This not only works to test Fourth World concepts, but it also works to remove unfavorable practices among objecting government bodies. Modern enterprises and Forth World entities alike should establish interoperability standards to aid in the economic transition required for Uploading. Here, users of any given service can easily switch to another, in a given territory, in order to allow individuals to move between new cities and become testers of new Uploading technology. Also, joining enterprises that align with Fourth World objectives can keep the system protected. For example, some telecommunication companies are currently exchanging certificates between one another to verify the source of a call and prevent call spoofing. This establishes a trust network that is impermeable to attack from those that don't align to Fourth World ideals.

---

science skill-set to form a specialized firm to perform mind mapping.

8. Open up global funding

If governments truly want to participate in innovation, their only choices are to provide patronage to Fourth World players and submit to joining forces with foreign investors. Here is where the intensely complex and intricate nature of today's investment community comes into play. Institutional Chinese investors such as SOHO are making investments in New York City. Japanese institutional investors such as SoftBank are taking large shares in European markets. They can bring returns from these investments back to their countries; at the same time, they can also block foreign investors from doing the same things. For example, China today, while investing externally, does not allow foreign investments internally. This causes competition between nations. Moreover, governmental interference is routine and is inversely proportional to investors' confidence in markets, so governments can prevent such investment from foreign bodies to benefit local corporations. But, this just causes more internal competition. Both forms of competition work against making big bets. For attracting investment bodies, governments will need to allow for a pooling of fiscal resources from a global investment community. Or else, The Fourth World will need to sideline the governments' parental role entirely.

9. Support pivoting funds from public and private investments.

This will require a shift in capital transfer through shrewd financialization and reinvestment of tax dollars that once served conventional public good directly. Here, society can collectively invest, along with institutional investors. Governments should ensure that regulations allow for easy reinvestment and VC raises that allow entities to make big bets.

As this model begins to mature, it enables those entities creating a tension with the Fourth World (i.e., those falling behind after creative destruction) to perish, complemented by a new socioeconomic structure that makes them:

- Unable to address consumer needs because they are already met through smarter, drastically more intelligent and efficient, systems
- Lack funding to transact within a system that makes big bets, as a trickle down of funds keeps true value held within tight tribal communities
- Fall short on pertinent social issues (e.g., being absent from discussions)

Ultimately, governments that don't evolve with the onset of a new social system — those that don't ensure that each entity has a distinct role in the Fourth World mission, don't regulate open integration between enterprises, and don't set standards for such exchanges — will fail. Older generations of governments will set laws that fall behind newer generations that understand how to use technology to break them.[52]

---

[52] For example, governments today provide access to transportation, but this will go away with mobility technology platforms that get funded by the investment community.

CHAPTER 8: NOTE

"The society based on production is only productive, not creative."

— Albert Camus

A new society vs. communism, socialism, and post-capitalism

While global powers meet to align, and governments restructure *what* they govern, making big bets requires single entities to control single assets to create a roadmap that delivers future human value. This approach, taken together with other concepts mentioned in this part of the book, may sound like communism, socialism, or perhaps post-capitalism. Here are the differences:

1. Unlike communism, economic resources are not controlled by the government. Like socialism, production capacity is communally owned. Unlike socialism, production is not managed by a democratically elected government. Like post-capitalism, citizens own production and distribution of resources. Unlike post-capitalism, there is no standard currency.

2. Technology will be leveraged in Unitism to establish a credit system, shared by a distinct community. Critics of socialism, and theories that don't utilize money, financial calculation and market pricing, state that models lacking a central

body to evaluate the worth of goods cannot exist. With Unitism, technology will serve to evaluate value.

3. Unlike post-capitalism, no highly-specialized organization will be owned by citizens to serve a community and no organization will have a shared currency. No organization will have shared credits either. Each individual's credits will be viewable on an open platform.

4. The Fourth World will have a variation of knowledge workers and service workers, as described in post-capitalist philosophies. Yet, labor will be specifically divided between those who leverage digital, physical, and biological technologies to capture a new type of value that prioritizes human evolution and those who serve one another to reshape our current human experiences.

5. Fourth World Entities are not concerned with the fair distribution of wealth. It is believed that communism failed in this vein, as the government never distributed wealth fairly because of intrinsic human selfishness. Distribution of wealth, in this scenario, will happen naturally through shrewd financialization, universal basic income (which can only be used to create goods for the community), and Gateway support of goods.

6. There is a fast failure system with Unitism. Monopolies can be created by commercial enterprise spin-offs, or any group of individuals forming a new entity. If these entities can't gain confidence that they are serving communities on a path to advancing humanity, they will cease to exist. Unlike the government's role in communism and socialism, failing entities will not be forced to reinvent themselves because they are tied to fulfilling consumer's subordinate needs.

7. There is a notion of duty for Fourth World Entities. This duty will be tracked and regulated.

PART III: MACHINES

CHAPTER 9: UPLOADING HUMANITY TAKES *COMMUNICATION WITH MACHINES*

"The best minds of my generation are thinking about how to make people click ads."

— Jeff Hammerbacher

The influence of a new collective

The reorganization of our social structure and the big bets we make on our future will clear a path for us to break the constraints that weaken us to act freely as individuals. We will no longer feel tied to the family constructs that are prevalent in our communities today. We will no longer feel the need to perform based on the financial success of the peers we grew up with. Our entire social design will be broken. It will allow us to be more tolerant of one another's differences to form new groupings. However, as we begin our journey on the Fourth World, and as we continue on it as Uploaders, our interactions will still be influenced by a third party system that connects all of us. This system is also a result of all of us.

While many of the materials that Fourth World Entities preserve will be used to advance the human physically, there will also be those entities that build a group of Distinct Intelligent Machines to support humanity.

Early versions of these machines are being born today. They are taking in a combination of all our thoughts. They are helping us accomplish rudimentary tasks in our day to day lives and enhancing our capabilities. As they help us achieve across various circumstances, they are having profound effects on our beings. They are influencing the parts of our brain that seek connection.

Feeding the machines a foundational culture, built on our collective thinking and sentiments, are networks that are constantly working to acquire data about us.

Over time, these machines will play a much stronger role in both our transition to a new material form and in navigating our future once we have fractionated off as Uploaders. They will remain connected to both the data transmission network and to us. They will replace the role that our economy plays today — they will provide services to support our lives with no exchange of value to the provider.

Eventually, they will become a crucial part of our survival. It is worth understanding them in depth.

What is Machine Learning?

Artificial Intelligence (AI) is a broad topic in computer science. The branch of AI, called **Machine Learning**, is what manifests itself most commonly in our human experience today. Machine Learning enables computers to learn on their own through a set of algorithms that parse data and discover patterns to make future decisions or predictions when faced with similar datasets. One step further is **Deep Learning**, which structures algorithms in layers to create **Artificial Neural Networks**.

The Artificial Neural Networks can be used to change one individual's sentiment, based on behaviors that other individuals exhibit within the network. This approach is already extremely common in personalized online advertising today.

Old-fashioned online advertising once served images to us based on the types of websites we went to. Computers were programmed to assume that everyone who visited a site was similar. But, in 1995, **Cookie** technology came along, allowing websites to grab a little more data about us. Eventually, advertisers picked up on this and began programming rules that slightly customized images to us based on these Cookies. But programming every rule into a machine was time-consuming. Difficult concepts would often need to be defined along the way, like, "show this image to this type of person."

Machine Learning, on the other hand, is programmed to teach *itself* by looking at a massive collection of data and finding its own patterns. Initially, patterns are found by connecting simple dots. But, with manual improvements in data classification, a programmer teaches the machine how to make more complex connections, perhaps even ones a human would never think of.[53]

Next, Machine Learning can be used to classify data. For example, it can look at a set of photographs and systematically classify backgrounds based on the presence of either smiling or angry people in the foreground of the photograph. Machine Learning can recognize that a group of young adults, who are all born in June, smile when they are at the beach, while another group of young adults, who were born in October, smile when they are in a mountainous region. This lesson can then be applied by presenting pictures of beaches or mountains to other young adults, not part of the original group analyzed. This method is used in advertising today, and it is the reason one individual may see a different image than their friend on a web-based travel ad. The machine knows which locations will appeal to the individual based on locations that it learned others who behave, act, or "look" like them online tend to enjoy.

Lastly, machines are taught to enhance and refine their learning, based on target outcomes. In advertising, the target outcomes typically orient on engagement with ads. So, if a young adult who was

---

53 Vincent, J. (2019, January 28). *The State of AI In 2019*. Retrieved from https://www.theverge.com

born in June doesn't click on a travel ad with a beach, the machine learns to either change the image for the individual or keep pushing the same image until the individual's psyche is changed to the point of positive engagement with the ad (i.e., the individual clicks it or, better yet, buys the service).

The question is, how does the machine know that the young adult was born in June?

The Cybernetic Collective

The example in the section above is very simple. As we are each complex beings with various attributes acting across multiple systems, we feed machines data that represents ourselves through various connected systems. Advertisement is the most robust driver of human to machine interaction at scale. However, there are other systems that collect other forms of data on us. Examples include those integrated with call centers to use our voice data for performing sentiment analysis and those that capture videos of us as we enter public spaces to do facial recognition.

The systems that collect human attributes have different components working together in the **Cybernetic Collective** (famously described by Elon Musk), and there are various levels of complexity within it.[54] As we feed in our collective data into a connected network of machines, we influence the culture of machines through Machine Learning and then AI influences us back much more aggressively. This cyclical manner highlights how data is collected on us and how we are currently being impacted by the data of all other individuals in our network:

- Platforms like Google, Twitter, Facebook, YouTube, and Amazon provide **Limbic Resonance** to consumers, stimulating the emotional part of the brain. Our nervous systems are not self-contained but rather attuned to those around

---

[54]Shead, S. (2018, September 7). *Elon Musk Smoked Marijuana And Chatted Cybernetic Collectives With Joe Rogan*. Retrieved from https://www.forbes.com

us. When we see content posted by those in our network, we begin to feel something, perhaps harmony, fear, anxiety or even anger.

- The cortex, our thinking part, just tries to make our limbic system, or feeling part, happy. So, when something that causes us to feel is contributed to the platform, our cortex drives a dialogue with those in our network to rationalize our feelings. That results in the contribution of even more data.

- Platforms are thereby able to accumulate a virtual treasure trove of verbal, audible, and visual data about us. They are able to use this data to personalize content for us.

- At one point, the communication with those in our network fails to stimulate a Limbic Resonance, at least to the degree that we need it, and we look to expand our networks.

- More data and more engagement typically result in more revenue opportunities from advertisers. To foster engagement, platforms typically utilize sensationalism, emphasizing the more limbic-stimulating content, provided by contributors, to help us expand dialogue by adding new members to our network.

- This economic necessity trains machines on the most excitable aspects of our culture, while it uses personalized content to train individuals to respond. So, we start seeing content from contributors that a machine thinks we *would* like. As we run out of options to satisfy our emotional responses, we eventually *do* like it. And, we begin to contribute more data, allowing the network of machines to learn more and more about us.

- Platforms that don't already have a lot of data about consumers are able to establish a deep understanding of us fairly quickly from data collected by aggregators. These aggregators create a unique identifier for each person across platforms, often bringing our Cookies together under one master identifier. Aggregators capture and sell data about the websites individuals visit, the offline purchases individuals make, and the geospatial locations individuals travel to.

- Any company can buy our unique identifiers and all the data attached to them. So, eventually, the unique identifiers that are collected for content personalization will work together to collectively program various aspects of Machine Learning.

- Since the network learns more and more about us, the data can be applied to influence our behavior outside of our limbic responses. Machines will be able to engage employees better, thereby, enhancing business operations (how should an individual be communicated with so that they perform their job better?). Machines will be able to provide better analytics for healthcare (will the individual get moody and act out publicly, or will they be depressed and exhibit inward behaviors as a response to a medication?). Machines will be able to figure out how autonomous vehicles should respond when they see a specific individual (is the individual prone to run from the car or stop?).

- Eventually, we will begin to trust these machines more across everything from religion to government. For example, in Japan, a Buddhist temple already has a robotic priest.[55] It uses AI to acquire more and more wisdom about faith. We can imagine it providing customized guidance to address an individual's issue. Or, as another example, the Estonian Ministry of Justice is developing a "robot judge" to adjudicate small claims disputes.[56] We can imagine a world where judgment is created based on our collective data, and we look to a machine to guide us on what is

---

55 Parke, C. (2019, August 4). *Robot priest added to 400-year-old Buddhist temple in Japan: 'It will grow in wisdom'*. Retrieved from https://www.foxnews.com

56 Niller, E. (2019, March 25). *Can AI Be a Fair Judge in Court? Estonia Thinks So.* Retrieved from https://www.wired.com

"fair." Or, in the case of the travel ad, we may one day see a robotic Psychiatrist that serves us a vacation ad in addition to prescribing medicine. The trust we instill in these systems will eventually further influence our behaviors.

Since Distinct Intelligent Machines in the Fourth World will operate with the data we provide it through the Cybernetic Collective, it's easy to see how a new artificial system will form as a response to today's collective human sentiment. It is also easy to see how Distinct Intelligent Machines will play a stronger role in our lives, and how the collective sentiment that they inherit will have a stronger influence on us.

Over time, these machines will all be connected to each other. They will form a new network and we will form a new Cybernetic Collective that engages through them. We can imagine a world where Galactic, Aquatic, and Mind Uploaders connect directly to a Distinct Intelligent Machine in a way that is, perhaps, more literal than how we connect to our computers or cell phones. They may take the form of Robots, and we could plug our human-machine-selves into them as the Distinct Intelligent Machines all connect to one another through a digital infrastructure.

Unequal data advantages

It should be understood that an unequal number of members have a say in influencing our Cybernetic Collective today. We can be sure that unequal influence will be placed on the new Cybernetic Collective.

Machine Learning-based products currently fight to get as much data from consumers as possible. But consumers only have so much data to give; they will only post their photos to a few social sharing platforms, will only be driving cars in so many intersections, and will only frequent certain websites. So, each AI-based product learns from a supply of data unequal to the amount of data every other machine learns from.

The data itself will not always be of equal value, either. And each contributor within each system feeds a different amount of data to the system.

How a system learns is also influenced by what it's taught to learn and how the programmer teaches it. Here too, there is an unequal advantage for programmers and enterprises that set target optimization goals. And then, at least in the short-term, there is an unequal economic response. Today, the effectiveness of AI products is usually measured by the business value their system produces. If the model produces perceived value, such as the ability to serve consumers with meaningful information, or if the model achieves needed business outcomes, such as the ability to serve advertisements, creators of the products will generate more cash flow to grow their capabilities. To fix this in a way that optimizes for future good, we can refer back to strategies for funding big bets to sustain development efforts.

Also, the cyclical connection formed between the consumer and the machine is heavily influenced. Advertising has a profound impact on consumers, yet it is pay-to-play; the richest entities simply pay more for more consumer mindshare, giving them direct access to feed data to and from the brain.

So, while a collective set of algorithms learns from the data we provide about our current society, a culture is created by the machines. That culture will be weighted in favor of select circumstances, individuals, groups, and platforms based on short-term economically valuable purposes. This, in turn, means that Distinct Intelligent Machines supporting Uploaders *could* be inappropriately influenced, unless today's society makes a conscious effort to sway its favor.

We have to understand that we all experience machines differently, but they all experience us in the same way. To them, we are just data, and Machine Learning needs a lot of data to acquire predictive power. Moreover, Machine Learning only knows what programmers teach them. And because machines don't *naturally* know what to do with the lessons they learn, they require additional engineering to make use of them.

Some machines — those that sit in the middle of active data collection, optimizing for target outcomes that are heavily funded by large enterprises today, and those that are programmed with the most advanced functions — will come forward with us as Uploaders. Some will become obsolete. And then, there are those machines that will be newly created, specifically to support Uploaders. These machines will connect to the longest-standing machines in the Cybernetic Collective when they need to access data to make decisions. So, how will the new machines know what to do in the Fourth World when they come across things we have never imagined today?

In the future, machines built specifically to support Uploaders will need to understand what kind of data to look for and the importance of specific data elements. They will need to learn things that we've never taught them to learn. They will need to take action on what they've learned. They will need to learn from their mistakes. They will need to know if their actions are correct or incorrect. We must begin anticipating these functions to shepherd this decision making.

Enter Artificial General Intelligence (AGI).

What is AGI?

Also known as "full AI," AGI refers to the ability of a machine to perform general intelligent actions, completing tasks that rely upon a full range of human cognitive abilities. While full AI isn't integrated into our lives today, it offers us a very important element for social transformation and future engagement with one another.

Full AI would be able to make decisions that balance multiple outcomes across multiple needs. They can optimize what is best for a human, considering all available known options. Unlike Machine Learning, for which code would have to be written to display an ad on a specific platform, AGI's decision-making capabilities could develop a strategy that makes choices across platforms.

For example, if a Distinct Intelligent Machine comes across a new organism, the Distinct Intelligent Machine could use Full AI to understand that we may need help approaching the new organism.

Using full AI, the Distinct Intelligent Machine could process the organism and tap into the Cybernetic Collective to determine that it is radioactive. It could use natural language to tell us if we need to put on additional protective gear when encountering the organism. Alternatively, the Distinct Intelligent Machine could recommend not engaging with the organism because it could cause harm. More importantly, it would be able to do so without the task of being explicitly programmed.

How would the Distinct Intelligent Machine create a strategy for cautioning us to put on protective gear instead of disengaging? Well, Deep Learning teaches the machines *how* to recognize things rather than *what* to recognize across multiple layers, as it marries up with **Reinforcement Learning** to create a responding strategy. Reinforcement Learning uses a feedback loop to train machines on how to determine the ideal behavior, within a specific context, to maximize performance and achieve a goal. The machine learns from interactions with the environment, rather than from being explicitly given ideal actions, by creating consequences of its actions through what is called a **Scalar Reward Signal**.

Determining which of the actions contributes to getting the reward requires the developer of the algorithm to set up a credit assignment problem. Actions are selected on the basis of past experiences as well as new choices. These choices are called states. In many cases, the machine is aware of its own states, or at least has some beliefs about them. Here, a technique called **Q-learning** is used in order to find an optimal action selection policy by modeling these states.[57]

In the Q-function:

- Q(s,a) represents the expected future reward when taking given action *a* in given state *s*.

---

[57] Given ever-expanding streams of data to navigate, there are, of course, much more complex programming methods than this, such as those that apply asynchronous strategies to learn from parallel copies of the environment

- The function also considers the long-term reward r subjected to some discounting factor y to account for the uncertainty of future actions and states.
- The algorithm works to maximize Q of taking action in state s to follow the best policy afterward. The formula looks something like this[58]:
- $Q(s,a) = r + γ(max(Q(s',a')))$.

And so, if the Distinct Intelligent Machine is optimized to perform on our safety, we see how full AI would allow engagement with many systems to make the optimal decision.

AGI vs. consciousness

The Distinct Intelligent Machine would not have the conscience necessary, like humans do, to make a judgment on whether approaching a new organism is morally right or wrong. This is because Machine Learning doesn't have organic intelligence and doesn't play by the same rules as humans.

Let's compare the learning of the Distinct Intelligent Machine to the life of a human. As a child, we are exposed to stimuli from the five senses: touch, smell, taste, sight, and hearing. Just like these programmed machines, we combine these experiences as states, or memories of the events and our responses to them. However, as humans, we each optimize in a very unique way. We don't know what the outcomes of our decisions are because they usually happen to us indirectly. We form connections that cross responses across unrelated events. At one point, our very complex mind decides to harden our response system in a very unique way.

Yet, machines can do facial recognition better than humans; they can even hear better than some humans. It is not inconceivable that they will also smell and touch better than humans. It is natural

---

58 Hwong, Y. (2017, November 7). *New Ideas in Reinforcement Learning.* Retrieved from https://agi.io

to wonder why machines can't convert their stimuli, actions, and responses into their very own consciousness. And such is the vision of "strong AI" that suggests machines can possess consciousness.

Berkeley Professor John Searle created "the Chinese room argument" to refute the grandiose claims of strong AI, claiming computer programs can only be formatted to simulate conscious states. In his experiment, there is a monolingual English speaker with a rule book that tells him what combinations of Chinese symbols to write when presented with other specific combinations of Chinese symbols. The English speaker acts as a computer, and the rulebook is his program.

Searle argues that the English speaker would be able to process the inputs and outputs perfectly, without having any real understanding of the meaning that the Chinese symbols represent. If the experiment was done in English, however, he would be aware of what was being said and the purpose it might serve. So, the computer program can process the syntax of a language, but that syntax alone cannot lead to semantic meaning in the way that strong-AI advocates hope.

Deep Learning and Reinforcement Learning are focused on optimizing machines to perform specific business outcomes, like enhancing the consumer experience or improving operational and financial performance. Machines aren't being programmed to perform the highly complex activity of the human brain; they are completely ignorant of the way a human brain really works. Products informed by ML are specialized, somewhat because it is easier to program a machine to do one thing than to do many things, but mostly because products are usually built with the intent of solving only one problem. Moreover, AI is incredibly inefficient in learning from its data. It needs to recycle the same information over and over again. On the other hand, the human mind is capable of learning from a limited quantity of data, otherwise known as a set of single experiences. So, each human experience is incredibly valuable. For example, if someone runs a red light and gets pulled over, they don't need to repeat that experience ten thousand times to know it's a bad thing to do.[59]

In the case of the new organism, once the Distinct Intelligent Machine warns us of the radioactivity, we may decide to tap into the Cybernetic Collective to seek more data about how to communicate with the radioactive organism. This is because we may have had one experience, early on in life, that showed how communication could overcome obstacles. We may "plug into" the Distinct Intelligent Machine and begin searching the new Cybernetic Collective for ideas on how to engage with the radioactive organism. And, we may apply the recommended actions from the Cybernetic Collective to help us make decisions.

We will discuss the importance of a connected system for our survival in later chapters. For now, it is important to know that we will grow in our connection and that humans will rely heavier on Distinct Intelligent Machines to feed data directly into their brains. Distinct Intelligent Machines will be connected to all other machines. An extended and more robust Cybernetic Collective is forming, and we must act consciously to feed it data today.

Stop being mean and click consciously

It is a reflection of our society today that people are drawn to cruel content. As far as the machines, well, they are just learning. They test responses to actions that they take and optimize them according to a business outcome. Optimizing for sensationalism was a test, and we failed as a society. We kept clicking the content with the most radical headlines. We spent more time reading the content underneath the radical headlines. We liked each other's bullying posts. So this content became the forefront of our searches and news feeds. We began to

---

59 *Artificial Intelligence Pioneer Says We Need to Start Over.* Retrieved from https://m-cacm.acm.org

think that that's what this world is all about. And now, it's making us mean. Meaner[60].

If we feed machines with mean data, they will become mean. If we respond to their meanness, they will become meaner.

But, most importantly, we must remember that what machines learn today will set a precedent for societies that Uploaders will live in as we evolve our species. And so, if we engage meanly with machines today, they will take the mean optimization models into the Fourth World, ruining the one opportunity we have to clear ourselves of all the meanness in this world, as we start anew.

We have an opportunity to alter the cycle. We can change how platforms influence us and how we, in turn, influence Distinct Intelligent Machines for the future. We can re-optimize algorithms.

Conscious human communication with machines is our best option for bettering humanity. In accordance with duty-based ethics in the Fourth World, society has a responsibility not only to ensure that Uploading is successfully achieved but to ensure that the world of Uploaders is better than the world we live in today. This involves being selective and careful about what it teaches machines. It involves making sure they have the "right" data to train themselves on. It means consumers should not click on content that angers them to fight the sensationalism that teaches machines that the new world should be dramatic. On the other hand, consumers should not shy away from transacting with data either. They should not disengage from data collection platforms out of fear of being known. Rather, consumers should give data thoughtfully and be mindful of how we transact with each other through machines, to set what is right and what is wrong today.

---

60 Machine intelligence is currently being optimized for meanness to achieve revenue dollars. This meanness will align with unique personal IDs and influence behaviors across the social ecosystem. And it will be too late to stop as machines will be smart enough to impact our behaviors. They will know us well enough through the data we feed them across system interactions.

Our world will transform from one where we are connected to machines into one where we are truly integrated with them. As this happens, Distinct Intelligent Machines will be responsible for connecting us, as a new human community, across Mind Uploaders, Aquatic Uploaders, and Galactic Uploaders. The question then remains, who is fit to enter this new network, and how will they be chosen?

CHAPTER 10: UPLOADING HUMANITY TAKES *SOCIAL SCORING*

"There are no facts, only interpretations."
— *Friedrich Nietzsche*

The effect begins

Advancements in AI will undoubtedly begin to impact our lives more and more. Notable Author and Futurist Ray Kurzweil discusses the notion of technological **Singularity**.[61] Defined as a point where technological growth becomes uncontrollable and irreversible, Singularity results in unfathomable changes to human civilization. Kurzweil claims once humans integrate with technology and Intelligent Machines, what he calls "non-biological intelligence," begins to dominate society. There will be radical changes in how humans learn, work, play, and even wage war. We will be able to metabolize food differently, have overabundant amounts of energy, and fight off infections much easier. We will be able to manifest our own form or select to evolve as we wish.

As we use technology to advance our experiences today, as we advance it to improve our health in the next few years, and as we adopt it more dramatically for pleasure, there will come a point of no return.

---

[61] First popularized by Vernor Vinge in his 1993 essay "The Coming Technological Singularity"

We will discuss this roadmap in the next chapter; however, now it is important to understand just how rapidly the effects of technology will come. And, it is important to know that the effects of Singularity will not spread out equally across society.

It could be conceived that as we segment ourselves into communities operated by Fourth World Entities, the experiments that are run on cohabiting groups will make them more inclined to a certain fraction of Uploading. Singularity will begin to evolve each city differently. And, it could be conceived that the Cybernetic Collective will impact us all to adopt certain common qualities that may actually not promote Uploaders.

Yet there is one thing we cannot argue: technology is already affecting all of us. Even if we are all kind and conscious in our engagement with machines, the effects of technology will undoubtedly continue to impact us all differently based on the geographic and socioeconomic conditions that we live in *today*. This is because access to technology is already distributed differently, and it is moving at different paces, benefiting some more than others.

For example, growth in mobile technology has not been equal across nations or within them. Today, people in advanced economies are more likely to have smartphones and are more likely to use the Internet and social media. Yet, within advanced economies, smartphone ownership can vary widely by country; 90% of South Koreans, Israelis, and Dutch people own smartphones, while only 60% of Russians own smartphones. And within those regions, younger, more highly educated, and wealthier people are more likely to be digitally connected.[62] Those who use smartphones for excessive typing are developing tendinitis in their thumbs. And so, it is not inconceivable that, in future generations, the thumbs of their children will evolve differently than the children's thumbs of the 55% of non-smartphone users in emerging economies today. Perhaps, as we form into cities where we serve one another, those with similar physical makeup will

---

62 Silver, L. (2019, February 5). *Smartphone Ownership Is Growing Rapidly Around the World, but Not Always Equally.* Retrieved from https://www.pewresearch.org

seek likeness in evolved adaptations rather than race or origin as we do today.

The minds of smartphone users are also evolving differently. On the one hand, personalized content helps these users learn more about their interests and to become more of the person they want to be. Those who are deeply connected may experience the effects of Singularity quicker. On the other hand, personalized notifications from mobile apps have already started to modify their prefrontal cortex. The prefrontal cortex normally deals with some of our highest-order cognitive functioning. Yet, the stress from the constant notifications has begun to establish a stress-fear memory pathway, making it go completely haywire. Those who experience drastic forms of stress induced by machines may shy away from them more and never get to experience their benefits.

As we begin to plan for a mass extinction event in the next 80 years, we must understand that we can each influence how Singularity will manifest itself. *Each and every one of us can have a say in our evolution towards Galactic, Aquatic, or Mind Uploading.* We will fraction off as separate species, but we don't have to fraction off based on the natural development of our physical forms today. We don't have to fraction off based on the benefits technology has already provided us as a result of our class. And we don't have to fraction off based on our desires to escape change. Instead, we can evaluate humans based on their **Fit For Circumstance** to break barriers and use that evaluation to help the right individuals integrate with the right technologies to prepare them for Uploading.

Social evaluation

It is important that a mechanism exists for evaluating humans. Just like machines are trained on what to learn to impact humans, we must use the power of social input to help us learn about ourselves and position ourselves for one of the Uploading paths in our evolution's destiny. We must recognize our true fit for evolving and surviving as one of the three Uploading species.

Today, we orient ourselves around our likeness with others and use grouping as a form of social survival. Take any lunchroom table at a secondary school. One who sits with the "athletes" orients on being a team player and survives through competition. One who sits with the "stoners" orients on being slightly rebellious and survives on being carefree. One who sits with the "nerds" orients on being academic and survives in thought-based pursuits. When one sits at one table long enough, they don't often switch to another table. But then, there are the very few that pride themselves on bouncing from table to table because they are "friends with everyone," "social butterflies," or, perhaps more truthfully, because they themselves don't know who they want to be. One thing happens at every secondary school though. There are always those that are not welcomed at any table, those that are banned. Those that sit alone.

The truth is, in life, we all get banned from something or another; we have all been rejected from a job, not invited to a social gathering, or lost a competition that makes us ineligible to join a team.

When humans feel socially rejected, the brain enacts a painkilling response by releasing opioids. The opioids trigger two areas of the brain: the amygdala, which processes the strength of the emotion, and the pregenual cingulate cortex, which determines how the mood changes as a result of the event. The response starts early and varies among people, with some brains releasing more opioids during social rejection than others, signifying that some of us have a stronger — or more adaptive — protective ability to fight against social rejection.[63]

Depending on the severity of the opioid release and the processing of the amygdala and pregenual cingulate cortex, when some individuals are banished, it can feel as strong as being physically beaten. Some may respond by becoming a recluse. For others, it ignites a belief that they have the right to expect ostracization, and they use it to enlarge themselves. They make themselves more visible to society. And then there are those that, perhaps, reflect on being ostracized to the point

---

63 Fisher, N. (2015, December 25). *Rejection And Physical Pain Are The Same To Your Brain*. Retrieved from https://www.forbes.com

where they could not handle the rejection, who would do anything so as not to be banned from their community as adults. All these responses are dangerous for evolution.

Historically speaking, banning someone, or being banned, has been a practice followed by communities as a way of enforcing a moral code, keeping members safe to engage in the practices of the community.[64] The mechanism has been employed for centuries as a way to protect honor. When humans rally around a code, they tend to crush anyone who goes against it. In the example of the lunchroom table, the "nerd" table may ban "athletes" because it threatens the growth of every other member's academic pursuits. It lowers the opportunity for the least "nerdiest" member to learn from "neardier" group members, and it lowers the average level of scholastic accomplishments. The "nerds" may, however, let in the "social butterfly" as this individual could potentially strengthen their growth. The new member could help socialize the academic causes. So, this is not to say that banning has been a way to keep new members out of a community. Rather, banishment has been a way of fighting for the vulnerable and, most importantly, a mechanism for survival.

Social evaluation through banishment is evolution's most natural form of selection. It is evolution's way of encouraging us to fork off in different directions based on the advantages each individual has to cultivate. And, we should listen to it.

Navigating community banishment is meaningful in the Fourth World on many fronts:

1. **Social Evaluation can help self-organize big bet community members.**

   Communities don't naturally orient around Fit For Circumstance for Uploading because most aren't aware of this option yet. However, most communities do understand when members are different from them, and this instinct can be used to reset

---

[64]*Invisibilia*. Retrieved from https://www.npr.org

their evolutionary path. It can optimize survival for Uploaders.

As individuals come to recognize the importance of Uploading, they can bond together and start to recognize how to maximize each member's chance of survival under each fraction. They can move, in groups, to new cities, where their theories can be tested with big bet technologies. Meanwhile, individuals who are banned from a group will be able to recognize that they may just be more fit for a different type of Uploading and this will increase the chance of survival for their lineage. Individuals may not always fit into the groups they *think* they belong in. And there is nothing wrong with that.

2. **Social Evaluation can help individuals join Fourth World Entities.**

    Those who act differently from most others in society may be fit for pioneering the Fourth World concepts. They may be built differently than others in a way that will let them withstand social rejection to shift energy in a positive direction, but more on this in later chapters. For now, it is important to know that some members who are often rejected from many groups may possess the skills to work on advancing Fourth World Entities. These members may be heavily recruited to advance Fourth World concepts.

3. **Social Evaluation can help individuals learn how to adjust.**

    Some of those who suffer social rejection may *not* actually represent the best in society. They may actually be harming the greater good. Not only should their data be used to train machines on bad examples, but these individuals should learn how to modify their behaviors. The important thing is not to force the direction society puts one in, but to use it to quickly learn how to enter into one of the Uploading paths, or to be a creator of Uploading as the possibility for humanity.

It should be clearly noted that banishment has nothing to do with social equality or access. It is simply a mechanism for oriening in survival groups. And, each group has equal value.

All in all, individuals should let the process of social evaluation serve them as a guide. It should help them set their path. *A method of social evaluation can lay the foundation for the creation of Founding Fathers of Aquatic, Galactic, and Mind Uploaders.*

However, social evaluations comes very rarely, and there are other signals that can guide Uploading groups. Enter the power of **Social Scoring**.

Social Scoring

A modern day system that allows for frequent analysis of individuals on multiple levels is scoring.

In Data Science terms, scoring refers to a prediction a machine learning model has on the outcome of an event. However, in social terms, this score simply tallies an individual's ability to exhibit a behavior within a certain defined category. It allows individuals to "vote" on other individual's attributes. Social Scoring simply enables classification.

Generally speaking, the ability to judge behavior is a skill that one hones over time. And, individuals are typically put in front of expert judges when they are either looking to be rewarded or acquitted of wrong-doing. In these cases, wrong-doing is tied to a set of predefined laws, rather than ethics. The government of China, on the other hand, is using a social credit system, akin to the Social Scoring, to judge every individual on their fitness to engage with certain activities based on their contribution to the greater good.[65]

---

[65] Ma, A. (2018, October 29). *China has started ranking citizens with a creepy 'social credit' system — here's what you can do wrong, and the embarrassing, demeaning ways*

Each citizen in China is ranked on the basis of a combination of all their social behavior, such as how well they drive, how much time they spend playing video games (seen as an unproductive behavior towards helping the greater good), how much money they waste on frivolous purchases, or if they smoke in non-smoking zones. A person's social credit can go up or down. Outcomes of this system, while serving ethical duty, can have negative impacts on citizens who, based on their score, can:

- Be restricted from taking a flight
- Be prevented from sending their children to certain schools
- Have their pets taken away
- Lose out on job opportunities

As the Fourth World evolves, each individual will have the opportunity to score one another.

Like the scoring system in China, Social Scoring in the Fourth World will look across multiple behaviors of an individual. Yet, unlike the scoring system in China, Social Scoring in the Fourth World will not punish anyone for low scores. Social Scoring in the Fourth world will use behavioral attributes to orient members to a community, motivate them to act in the best interest of the community, and segment individuals based on their fit for a fraction of Uploading.

Social Scoring in the Fourth World will also remain free from the government's involvement. As entities forming in the Fourth World will deliver on the blueprint for Uploading, and as excess funds of today will shift to sustain communities, a renaissance will develop to fulfill one another's wants. New, unregulated, untaxed currencies that create true value will shift power away from governments and place more responsibility on community members in new cities. One of these responsibilities will be scoring members within the community. Moreover, Social Scoring will be a requirement for joining new cities.

The new system will allow members to regulate one another for community fitness. As new cities form, Social Scoring won't

---

they can punish you. Retrieved from https://www.businessinsider.com

necessarily represent good or bad in universal terms. Social norms will exist within each community. Each community's norms will be different but aligned to one of the three fractions of Uploading. No single party or individual will decide the norms. Rather, groups of people will align on them and bond together because of them. Those that don't fit community norms will be encouraged to join other communities that align with their values. Scores will be visible across communities. Preexisting scores from one community can be used to easily accept new members into another community.

Individuals will *want* to be scored as entry into a Fourth World community provides benefits, such as access to Universal Basic Income, free use of community resources, and participation in a currency that represents true value.

Social Scoring will also be used to help members within a community represent kinder versions of themselves and build a social muscle capable of surviving outside of society as we know it today. To get good individual scores in certain areas, people will want to provide good things to one another in their communities. They will want to help each other.

Scoring will be weighted. "Votes" from individuals with the highest scores in one area will be valued more when assessing other individuals in the same category. This weighted form of judgment can also help machines understand whether or not the data they are gathering from individuals should be weighted as good, based on the human-prescribed characteristics of the individual. Machines can apply their own weight to this data to help us control how we influence one another in the Cybernetic Collective. If machines draw intelligence from humanity in this way over time, they will survive the regeneration of the planet with a culture that best represents the most direct intents of today's social environment. Machines can also use less data to learn more, faster.

Ultimately, though, Social Scoring can help society decide on how to fraction the new branches of humanity — who to transform for Uploading into the galaxy, who to transform to stay on Earth in the Ocean Atlantis, and who to transform to become a Mind Uploader. It can ensure that those who are selected to branch off are truly fit for the new circumstances they will encounter.

Social Scoring will entail an element of classification that can help predict the success of human survival in future circumstances. Fourth World Entities can use the information to customize technology that will advance each individual's capabilities. It can help us understand what type of people bond together most strongly for survival.

For example, some minds may be better suited for eternal survival, and they will take the form of Mind Uploaders. These are the minds that will be desired for study for years to come. They will be scored highly for their wisdom. Other minds may be better suited for exploration, and they will take the form of Galactic Uploaders. They will be scored highly for their curiosity, ability to collaborate, and cleverness. Then, of course, there are those humans who are better suited for community evolution thanks to their kindness, tenaciousness, and nurturing nature. The Ocean Atlantis presents itself best for this type of scored individual.

More importantly, scoring can help ensure that those who engage in hate crimes and mass murder events, such as the ones we see so commonly in the news, do not make it into the Fourth World. It can ensure the purity of the Fourth World. It can promise a true reset; those who are not scored for fitness with any community do not get the benefits of integration with Uploading technology. It preserves a kind culture.

CHAPTER 11: UPLOADING HUMANITY TAKES *NEW DECISIONS — AQUATIC UPLOADERS*

"Far and away the best prize that life has to offer is the chance to work hard at work worth doing."

— Theodore Roosevelt

The Ocean Atlantis

Aquatic Uploaders will form sets of colonies in Ocean Atlantis. Underwater domes will be dispersed like freckles on the body. They will be grouped together, some domes bigger than others. Most domes will be big enough to fit a 10-20 families. Some will fit over 40 families.

Aquatic Uploaders will live off the land. They will use hydroponics to create fresh water through desalination.[66] They will grow plants in pods on the seabed and survive on the nutrients of fresh fruit and vegetables.[67] They will augment their diet with food-pills or feeding tubes.

---

66 A process that takes away mineral components from saline water

67 McEachran, R. (2015, August 13). *Under the sea: the underwater farms growing basil, strawberries and lettuce.* Retrieved from https://www.theguardian.com

The physical form of Aquatic Uploaders will resemble the humans we see today, more than the other two forms of Uploaders. Over time, they will naturally adapt their eyesight to see perfectly underwater. They will evolve to be hairless, with webbed feet, for easier navigation in the water. Aquatic Uploaders will also be fitted with new materials. Perhaps, their cyborg qualities will involve integration with rebreathers (devices that permit the recycling of unused oxygen content of each breath). Oxygen stations will replenish the amount metabolized by the Uploader. It will be the new oil, generating energy for individuals.

They will procreate similarly to the human of today; however, their domes will hold tanks of DNA cold storage, and they will be able to augment and alter their kin. It will involve DNA grafting, similar to the system we see emerging with CRISPR today (i.e., the ability to edit genomes, allowing researchers to alter DNA sequences and modify gene function easily).

Since domes will only be able to hold a finite number of members, Aquatic Uploaders will spend most of their days building new domes and resettling new communities. Each dome will have a slightly unique culture — members will form tribes.

Aquatic Uploaders will be supported by Distinct Intelligent Machines — underwater drones and robots — that will assist them in their exploration and recreation of new domes. The Distinct Intelligent Machines will be brought with Aquatic Uploaders in initial waves of migration underwater to escape the mass extinction events. New machines will be created once these extinction events occur.

Aquatic Uploaders will transform elements that they find underwater — those, such as plastic, that we leave behind today and consider waste or objects that will be washed into the water after the mass extinction event — into purposeful goods and building materials for new domes. Vessels will transport Aquatic Uploaders between dome shelters to exchange underwater farming practices, dome creation strategies, and physical objects.

Breakthroughs & advancements

There's certainly progress being made to advance the path for Aquatic Uploaders by building an Ocean Atlantis. In 1962, diving pioneer Jacques-Yves Cousteau built the first underwater habitat. He lived in it for 30 days, submerged off the coast of Marseilles, as he built an underwater farm. Years later, in 2014, Cousteau's grandson, Fabien Cousteau, managed to live 31 days in the undersea laboratory called Aquarius, outside the Florida Keys. Aquarius was an 81-ton vessel equipped with hot water, a kitchen, air conditioning, and computers. While Fabien Cousteau took his grandfather's mission of ocean exploration further, setting a new record for the longest time spent underwater, Canadian inventor Phil Nuytten took underwater farming further when he released a design for a settlement deep underwater. The settlement would harvest resources from thermal vents. Meanwhile, French Architect Jacques Rougerie, developed the first component of the SeaOrbiter, a floating research colony that extends thirty-one meters beneath the water surface.[68]

Aquatic Uploaders will use the advancements of the early pioneers to colonize the waters of the Earth. As they grow in numbers, they will establish a unique way of being. To get a glimpse into their way of being and to understand how individuals today should be scored to fit the future circumstances of Aquatic Uploaders, it is worth looking at a group of people living in Peru who have formed tribes above the water's surface for generations.

The Uru

Lake Titicaca, nested in the Andes on the border of Bolivia and Peru, is inhabited by the Uru people. Here, they created floating islands of handmade, dried, bundles of reeds. Legend has it that the Uru people considered themselves to be the owners of lakes and waters. They thought themselves to have black blood because they did not feel pain

---

68 *Could humans live in underwater cities?* (2018, November 21). Retrieved from https://www.sciencefocus.com

from the cold, and in the Andes, it gets *cold*. To escape slavery and taxation by the Inca tribes, they created these islands; if a threat arose, the islands could be moved.[69] The Urus were eventually still conquered by the Incas in the 1400s, but they remain on the islands to this day.

Their island settlements are small, only a few homes per island. The homes themselves are made from the same dried reeds used to make the islands. They are humble, with only a single bed for sleeping, shared by the entire family. When families fight, they simply split the islands; when families form, they join islands together. They generally work to keep peace in their community so as not to labor over island reconstruction.

Meat cooks on their roofs in the sunlight. They harvest fish and move from island to island on beautifully decorated man-made boats; beauty is important to them. Toddlers are bound to poles, with a string, so that they do not fall off the island edges. But many still do. In fact, amidst the cold and heat inherent in their topography, children falling off reed islands is the biggest cause of death for the Uru people. Yet, as a people, they survive. Looking at the Uru people, one would find the qualities necessary for successful Aquatic Uploaders.

Scoring for Aquatic Uploaders

Modeling on the Urus, humans who are scored with the following characteristics will lend themselves well to Aquatic Uploading:

- **Amenable.** Harmony in underwater domes will be necessary as Aquatic Uploaders will live in small communities. As such, most will be kind and harmonious.
- **Good designers.** As Aquatic Uploaders will make use of human DNA to procreate. They will also be able to perfect the human form to rid itself of weaknesses.

---

69 Wikipedia contributors. (2019, September 25). Uru people. In Wikipedia, The Free Encyclopedia. Retrieved 22:03, October 6, 2019, from https://en.wikipedia.org

- **Aptitude to perform manual labor.** Although all Uploaders will be fit, as cyborgs, with machine-enhancing and life-prolonging technology, Aquatic Uploaders will be the least transformed, and so they will need to possess the ability to perform hard, manual labor. They will be tenacious, nurturing, and hard working.

CHAPTER 12: **UPLOADING HUMANITY TAKES** *NEW DECISIONS — GALACTIC UPLOADERS*

"Space exploration is a force of nature unto itself that no other force in society can rival."

— Neil deGrasse Tyson

New planets

Waves of Galactic Uploaders will venture off to new worlds. These Uploaders will value novelty, expansion of the mind, and freedom. Sure, they will be cyborgs with advanced capabilities, but they will still have aspects of the human that we experience today: a genetic desire to seek stimulation, a strong response from the surge of dopamine and adrenaline released when they have a novel experience, both of which combine to create a sensation of being truly "alive". So, they will be driven to travel to new planets, believing that aliveness can ultimately be found in never-ending capacity.

Galactic Uploaders will be augmented with heavy technology as they venture off into space. They will not all be launched into space at once. And so, over the course of the next 80 years (prior to the mass extinction event), each subsequent launch will enable Galactic Uploaders with deeper and more robust technological integrations.

The first wave of early space colonists will be augmented to create an early settlement. Perhaps integration with machines will impact their skin, enhancing it so that they are less affected by the harmful rays of local stars but also impacting their touch receptors to change how their brain receives information about their environment. Perhaps, they will have adrenal gland implants that calm the release of their adrenaline so that they respond rationally when approaching danger.

The first wave of Galactic Uploaders will send feedback on their augmentation back to Earth, which will inform advancements for new cycles of Uploaders. In fact, these communication systems will continue to exist for Galactic Uploaders who will communicate their learnings to Aquatic and Mind Uploaders throughout the next wave of our evolution, but more on this in later chapters. Over time, Galactic Uploaders will begin to augment themselves with materials and new technologies that are unavailable to biomedical scientists on Earth. Eventually, as cyborgs, Galactic Uploaders will vary drastically, as one new species, in their physical makeup.

Unlike Aquatic Uploaders, not all Galactic Uploaders will form communities on the same planet. A community will form on each planet and be used partially for full inhabitants and partially as hoteling stations for other Galactic Uploaders.

They may colonize Mars first. Unlike our Moon, which has drastic temperature changes of about 572 degrees Fahrenheit between day and night, the atmosphere on Mars has carbon dioxide, which allows wind to blow and helps to equalize the day-to-night temperature differences. The atmosphere also allows us to pressurize domes and structures using air from outside, and it provides conditions for plants to grow. Also, unlike the days on our Moon, which are equivalent to 28 Earth days, Mars is just a little over 24 hours long which is far easier for humans to adapt to. Mars also has a stronger gravitational force than the Moon, signs of water, and better UV protection (if we lived on the Moon, we'd have to live in caves to protect ourselves from getting obliterated by the sun).[70]

Breakthroughs & advancements

NASA *has* done a little to sustain life in new environments. It has already considered the kind of habitation we'll need to survive on Mars — a self-sustaining environment, sealed against the thin atmosphere, equipped with life support systems (ECLSS), power systems, renewable resources, and docking ports. The research gives us a glimpse into the life of the environment that some Galactic Uploaders will settle into. From Mars, other Galactic Uploaders will venture even further beyond. They will travel in small vessels, and they will be equipped with even fewer, although more intelligent, tools that those on Mars to help guide them. Just like Aquatic Uploaders, they will have help from Intelligent Machines. However, these machines will run very different models to the ones used by Aquatic Uploaders, as they will be built to help Galactic Uploaders make decisions when they explore worlds that we have yet to imagine.[71]

The initial waves of Galactic Uploaders will be quite small in quantity. Few, mostly because the time, cost, and capacity preclude more than a few today. When Earth and Mars are closest to each other, the trip there will take 260 days. As one of the earlier commercial space pioneers, Jeff Bezos, himself, is already taking payments of $200,000 – $300,000 for future trips on his yet-to-be fully developed rockets to space, and that's just for a few minutes' ride into orbit. But

---

70 Ridley, A. (2018, October 18). *Is it better to live on the moon or on Mars? A scientific investigation*. Retrieved from https://qz.com

71 A Dynamic Bayesian Network is a mathematical model that relates variables to each other over adjacent time steps. It says that at any point in time, the value of a variable can be calculated from immediate prior values. Yet there are unobservable hidden states, and these can be modeled using the Viterbi algorithm that determines the most likely sequence of events to find a sequence of observed events. These models are used today in everything from cellular signals to bioinformatics to robotics, speech recognition, digital forensics, and protein sequencing.

that rocket only has six seats. It's an understandable price given the costs of space travel. Elon Musk, owner of SpaceX, says that it costs $62 million every time its Falcon 9 rocket is launched, while the more powerful Falcon Heavy costs $90 million per launch.[72] So, there aren't many rocket launches.

Given the grand Galaxy and the few Uploaders that will venture off to it, we can look at the life of the Bedouin, living on Earth today, to get a sense of what life will be like.

The Bedouin

A week in a small village near Jordan's Wadi Rum desert and Zaid was done. Too much sweet tea from his Bedouin uncles, excited to see his shining bright eyes. Too warm from sleeping in Mickey Mouse blankets tucking him into cot beds. Too full from canned tuna, stale pita bread, goat cheese, and cucumbers. Zaid was ready to leave his base station and return to his nomadic lifestyle.

Ziad is one of many Bedouins roaming the orange rolling sand hills of Wadi Rum. He takes a beat up old Toyota Land Cruiser deep into the desert. The rust breaking away almost all of the white paint doesn't bother him. Ziad stops near a group of rocks and spends his days exploring ancient carvings, wondering how different life was for the nomads that carved them. As the sun begins to set, he finds some sticks and rubs them together to make a fire and cook the food he brought with him from the village.

The aroma of burnt wood and spiced goat meat fills the air. He is joined by other Bedouins who bring their camels. They smell the food from far away.

At night, Zaid sleeps uncovered in caves under the stars, disconnected from a world that values time and money over peace. Zaid could have left his family village, when he was younger, and chosen to work in the neighboring city of Aqaba. Yet, to his somewhat debatable

---

72 Tuttle, B. (2018, February 6). *Here's How Much It Costs for Elon Musk to Launch a SpaceX Rocket*. Retrieved from http://money.com

fortune, he was built for curiosity and simplicity. Even if he tried to go to Aqaba, he knew that the strains of work pressure would yank that state of settlement right out of our DNA.

When Zaid wakes up he analyzes the gas that he has left in his Land Cruiser. It if is enough to go further in the desert and still return to the village safely he continues. Even if it isn't enough, he may choose to leave his Toyota in one spot and keep walking. Eventually, he returns to his village with experiences to marvel the minds of his fellow Bedouins. He is satisfied with his bragging rights.

Life for Galactic Uploaders will be similar. They will travel too and from base stations on neighboring planets. There, they will replenish, gather new supplies, and continue venturing for the sheer sake of exploration.

Scoring for Galactic Uploaders

Even though technological advancements are moving quicker than ever before, and entities can make progress in creating more rockets at lower costs, society will still need to be judicious on who it will select as early pioneers of Galactic Uploading. Selected individuals will need to be fit with characteristics to not only survive in new environments but to develop new offshoots of Galactic Uploaders. Characteristics for Galactic Uploaders can begin to be developed by humans today. The following qualities can ripen, over time, to evolve Galactic Uploading in the future:

- **Ingenuity.** Galactic Uploaders will need to invent things that they have never seen before in order to survive. They will have to invent new breeds, even within their own species to fit into new planets. They will be selected for being naturally clever.
- **Resilience.** Life will be hard for Galactic Uploaders. They will have to survive anything that comes their way without allowing it to break their psychological makeup. They will need to constantly collaborate with others to ensure their survival.

- **Curiosity.** Galactic Uploaders will need to have a natural desire to explore, or else the very function their species serves will fail.

- **Peaceful:** Galactic Uploaders will experience long bouts of science. The journey between planets will be long. They will need to have an innate stillness to embark on each new journey.

CHAPTER 13: UPLOADING HUMANITY TAKES *NEW DECISIONS — MIND UPLOADERS*

"The energy of the mind is the essence of life."
— Aristotle

On land

Mind Uploaders are, perhaps, one of the most important fractions of the future human. These Uploaders can keep humanity surviving in a *totally* new material form — one that is made up of machine, alone.

Mind Uploaders will remain on Earth, but they will require different conditions than those venturing into space or underwater. They will not be able to move about as freely or easily as Aquatic and Galactic Uploaders. Mind Uploaders will rest in computers, protected from environmental conditions. They will not be impacted by famine caused by drought and deforestation because they won't depend on nutrition or oxygen for survival. They will be protected from volcanic eruptions, meteors, and other cataclysmic natural disasters because they will be backed up and distributed in durable data centers across the Earth. Mind Uploaders will also need a heightened level of digital security placed on their preservation. These minds can be hacked. False memories can be inserted into their program, threatening humanity's time capsules.

Although Mind Uploaders are the last fraction of Uploaders to mature, they will eventually serve to guide future Galactic and Aquatic Uploaders. Unlike Galactic and Aquatic Uploaders, who will need to "plug into" Distinct Intelligent Machines to access the Cybernetic Collective, Mind Uploaders will be constantly connected to the new Web. They will have the ability to synthesize information, and they may be drawn on by other Uploaders for advice on how to make decisions in new environments.

With Mind Uploading, we can imagine the mind of Albert Einstein, perfectly captured in a computer system, engaging with the current generation, as he would have at the peak of his human state. What lessons died with him that could be brought forward? What would he learn from us, and how might he use what he learned to make his mind even stronger?

With Mind Uploading, we can imagine a world where the wisdom of the old is not only passed onto the new but where it is also capable of transcending generations to apply itself to the new. A world where, with the extension of its existence and the preservation of its strength, the brain keeps building on itself and keeps learning from its mistakes.

With Mind Uploading, we can imagine a world where, like cyborgs, the brains of those preserved forever can be altered. Memories that hold one back are zapped out, suppressed. Memories that open up one's world to positive responses are brought to the surface. Existence for those minds, Uploaded, would contain no pain. They would exist in a constant state of pleasure.

Breakthroughs & advancements

It is worth understanding how the evolution of Mind Uploading will take this form. Mind Uploading requires scanning and mapping the features of the brain and then transferring them for storage to a computer system that runs a simulation of the brain's information processing, such that it responds in essentially the same way as the original brain and experiences a conscious mind. The mapped cognitive processing can be implemented on substrates within or external to the human body. Doing so allows humans, through their consciousness, to transcend their current physical experiences.

Multiple technologies that approach this actuality are currently being developed. Generally speaking, they range along a continuum from neural prosthetics to Whole Brain Emulation (WBE). With these technologies, portions of the brain will be gradually replaced with neuroprosthetics. For example, the part of the brain that stores memories may be replaced with computer parts and be connected to a network of other computers with wires, or perhaps even wirelessly. WBE, on the other hand, requires humans to die for full functionality. When the human is alive, brain activity can be monitored and recorded. Then, upon death, the biological brain can be embalmed and thinly sliced so that neural connections can be mapped and reconstructed to form a working artificial brain with information that connects the slices using previously recorded data. Today, brain cells begin to die in less than five minutes after their oxygen supply is cut off. So, scientists believe that humans must "die" before WBE can begin. Technology may alter this approach over time.

No, this isn't AI. WBE technology is programmed differently from Distinct Intelligent Machines. Although it leverages large-scale datasets, it has a consciousness. Nor are these cyborgs. These brains are not augmented to see better or hear better. These are imperishable capsules experiencing new experiences as though they were their own, consuming more information than ever thought possible by humans today and responding to it in programmatically untrained ways. These are actual human minds navigating a digital and physical world.

The Kasbah of Telouet

The Kasbah of Telouet sits 4 hours outside the city of Marrakesh down a windy road. On the way there, bursts of small Berber towns peak out through the red mountains separating themselves from lush greenery in the background. Shepherds herd goats and sheep. Women carry heavy grass in baskets half their size on their backs. Little kids play with balls amidst running chickens and donkeys. The journey there takes its way through Tizi-n-Tichka mountain pass, the highest in the Atlas, standing at 2,500 metres. Arriving at the town surrounding The Kasbah of Telouet castle, a few small clay houses hold no more than 300 people.

In the distance a windy trail leads deeper into the wonder that is Africa. In the late 1800s, camel caravans took the very same trail, passing the patch of grass that the Kasbah of Telouet hoists on today. They came from Africa and Asia carrying goods and slaves to trade from their homeland.

The Kasbah of Telouet was once the home of Pasha Glaoui. He had a strong knowledge of the resources that were inherent to his region, namely the salt that came from the surrounding mountains.

Glaoui was born to a concubine and local baron of the mountainous region. One day, the Sultan Hassan I was passing through the trail on a tax-gathering expedition. A blizzard left the Sultan stuck in the mountains, starving. Glaoui was passing by and saved the Sultan from death. As a reward, the Sulton granted Glaoui possession of a cannon, which was then used to subdue rival warlords in the area.[73]

Glaoui used the cannon to trade the salt with travelers that passed. Salt was used to preserve food, and so its value allowed Glaoui to amass bricks of gold and silk. The salt became so popular that caravans would come in pursuit of it. To this day, one of the largest Moroccan industries is packaging and preservation of canned goods.

Pasha Glaoui used his gold and silk to build the The Kasbah of Telouet. The next two generations of Pasha's sons began to cause strife. Amassing 1300 slaves, the second Pasha enhanced the castle with even more wealth. He treated his slaves poorly and many died.

Prior to the second Pasha's position, the French and British had several attempts at dominating Africa, but despite their interest, Morocco was able to maintain independence. In the latter part of

---

73 (2010, 28 June). *Reviving the last Pasha of Marrakech*. Retrieved from http://news.bbc.co.uk

the 19th century Morocco's instability resulted in European countries intervening to demand economic concessions. The Pasha sided with the encroaching French at the start of the 20th Century as they colonized Morocco. As a result, he was rewarded with a sizable slice of the region's economy through trade of olives, saffron, salt, mineral mines. He was also given a cut of the income generated from Marrakech's red light district. This angered the people of Morocco.

But the third Pasha Glaoui made relations with both the Moroccan people and the French that colonized it even worse. By this time, now in the 1940s, the third Pasha was considered the most powerful man in Morocco and one of the wealthiest men in the world. He had 5 wives, while the Koran, as practiced by people in Marrakesh only believe a man could have 4 wives, and only if he has the wealth to treat each one of them accurately. He also had 60 concubines and acquired hundreds more slaves who he enlisted to build the Kasbah up even grander. Looking to strengthen his position globally in the 1950's, the Pasha Glaoui went to the queen of England looking to get Knighted. After being refused, he went back to his proceeding father's alliay and helped the French retain their position against the rebelling Mohamed V and the people of Morocco who were organizing an independence movement. As a result, the French wanted the Pasha to rule the nation.

Meanwhile, Mohamed went into exile. However, the French realized that Morocco was descending into chaos and granted it independence in 1956. The Pasha Glaoui was considered a traitor for siding with the French colonists and helping them maintain power over Morocco. Mohamed V seized his property and his Kasbah fell into disrepair as Paha eventually died in 1956, a broken man.

The Kasbah is not protected by UNESCO and remains refurbished by the Morrocan government because of the betrayal by Pasha's family.

The story of Kasbah of Telouet shows us how a great asset can preserve the history of the people. In the Fourth World, the asset is human. It is the Mind Uploader. But this story also shows how, if unprotected, it can be destroyed. Therefore, Mind Uploaders will need to be selected carefully and protected from corruption at all costs.

Scoring for Mind Uploaders

Since the most valuable information will come from Mind Uploaders, they will hold the wisdom revered by fractions of other Uploaders. As such, Mind Uploaders will be scored with the following characteristics:

- **Elevated IQ.** Although the new world will be good because we will have designed it this way, Mind Uploaders will need to learn quickly in case there are bad actors to protect themselves from a hack or other mental interference. And although integration with technology will advance all Uploader's learning, Mind Uploaders will need to leverage the best of human minds today to help other Uploaders make educated decisions to inform their survival techniques.

- **Independent** Mind Uploaders will live in solitude. They will not procreate. They must find a peaceful freedom in existing within themselves.

- **Insightful.** Mind Uploaders will be a protected class. They must use the parts of humanity with the most sagacity to navigate all the options of the digital world. However, this doesn't mean that they need to understand the digital world and all its workings. In fact, many Mind Uploaders will be indigenous people with few skills in engineering or technology.

We can conceive a transitional path where humans, perhaps as cyborgs, integrate portions of their brains to communicate with technology that enables Mind Uploading. Prior to partial integration, it is inevitable that a few people will sacrifice their brains to science for this possibility. Some people will volunteer to die for duty and the betterment of others. Those who make this sacrifice will be given solace as their minds could be manipulated to find joy as they fall dead to the regenerating planet. While they will not truly emerge as Mind Uploaders of the Fourth World, we must still recognize them as early Mind Uploaders. For both the early and the eventual Mind Upliders, we must understand that there can very well be a world of cognitive utopia for the last remaining full-fleshed humans that enter this path. To understand their path, and the path of all other Uploaders, it is worth looking at the roadmap for physical integration in the Fourth World.

CHAPTER 14: UPLOADING HUMANITY TAKES *PHYSICAL INTEGRATION*

"The more you extend your senses, the more that you realize exists. If you're in the same house for years, there's a repetition of what you perceive there. If you add a new sense, though, the house becomes new again."

— The first cyborg, Neil Harbisson

A roadmap for integration

The path to each fraction of Uploading, within each community surrounding a Fourth World Entity, will take place in stages.

As we accept that humanity evolves through both technology and socioeconomics and if we marry that with the intent of Social Scoring in the Fourth World, we can begin to see how we are producing a path forward for humanity:

```
Data collection systems → Data processing systems
  → Data storage systems → Data learning systems →
 Machine-to-human interfaces → Biological scanning,
genetic mutation, and prosthetics → Social Scoring →
     Do-it-yourself evolution towards Uploading[74]
```

---

[74] *Beyond Human.* Retrieved from https://www.nationalgeographic.com

Technological advancements are still limited; however, they hold the potential to accomplish extraordinary feats. The science used to advance one technology, which makes our way of life easier today, can be leveraged to catapult technologies in the Fourth World.

Interacting with machines to enhance our experience

We already have a superficial connection to machines in the sense that we use them to communicate with each other in new ways. This interaction strengthens the power of the machines while the economics of their engagement with us provides humans with stimulation.

And already, we see humans existing very emotionally within a culture of machines. They fall in love with them.

In November 2018, Akihiko Kondo, a 35 year old school administrator in Tokyo, walked down the aisle in a white tuxedo. The bride, Hatsune Miku, was not only a world-famous recording artist throughout Japan, she was also a hologram. After years of feeling ostracized by real-life women for being an anime geek, Kondo felt the wedding was a triumph of true love. He considers himself a sexual minority facing discrimination. Yet, he is just one of the first to enter, very publicly, a second wave of what is called "digisexuality". The first wave is experienced through online pornography, hookup apps, sexting, and electronic sex toys. Here technology is simply a delivery system for sexual fulfillment and leaves openings for physical connection with other humans. The second wave of digisexuality, however, obviates the need for a human partner altogether; it allows humans to form deeper relationships through immersive technologies like virtual and augmented reality.[75]

Although this example is fairly extreme, we can't deny that many individuals are forming a deep bond with their connected devices. The Cybernetic Collective of today will stimulate us to engage more,

---

75 Williams, A. (2019, January 19). *Do you take this Robot?* Retrieved from https://www.nytimes.com

but AI will make the bond we have with the machines that we use to engage with feel even more personal. We can imagine connecting to AI-based machines that will supply us with the value we seek from human friendship today. The machines can serve as physical and intelligent companions. We can imagine machines replacing physiatrists or social workers, maintaining the power to gain our affection by storing all our memories and playing them back in front of us when we want to just "watch" them. We can begin to care more for objects as they teach us and inform us.

But, most of us still need to sit in front of a device, be it a computer, smartphone, camera, etc., and perform human actions in order to experience any kind of stimulation from machines.

Commingling with machines to improve our health

A rather seamless human-machine interface has already started emerging in the form of machines being integrated into our own body as prosthetics, microchips, or implants. These items can relay information between the body, the mind, and external entities, such as other computers and people (think health monitoring or regulating systems measuring and communicating the levels or rates of processes in the body). And they do so at speeds that are billions of times faster than the processing time of the current human-to-machine interfaces.

Artificial replacements are currently available for many complex organs, as well, including the ears, eyes, liver, lungs, ovaries, and pancreas. While some of these are more mechanical, only requiring moving parts and an internal power supply to keep them moving (e.g., the pacemaker), others are much more complicated, such as the nose and eyes that communicate with the brain to send accurate signals regarding the state of the world outside of their host body. These machine parts can be connected directly to the brain in order to allow us to control these advanced replacements. In order to function, these replacements require not only to be physically installed but also to be literally integrated with the brain in order to properly send signals of sight and smell.

The brain itself has already begun to see artificial replacements. Neural prosthesis replaces parts that have been damaged by disease or injury to enable motor, cognitive, and sensory functions of the brain. Neurostimulators and deep brain stimulators send electrical impulses to the brain to treat a variety of disorders such as Parkinson's, Epilepsy, Urinary Incontinence, and even treatment-resistant depression.

These machines will further our roadmap of integration as we can use them to enhance the way we experience the world. Over time, just as we love the external machines built with AI, we will begin to love the parts of our bodies that enhance our health.

Human augmentation to improve our sensing

We have begun to embrace human augmentation beyond medical aid to creatively experience the world in a totally different way. English Cyborg Artist Neil Harbisson was born with complete color-blindness.[76] Yet, for the last thirteen years, he has been able to "hear color" through an antenna-like sensor implanted in his head. The sensor translates invisible wavelengths of light (i.e., the infrared and ultraviolet color spectrums) into vibrations on his skull, which he then perceives as sound. It is an example of a human existing with new wiring.

This has transformed Harbisson's life in ways unimaginable to the non-cybernetic human. For example, he can now understand which days the sun has the most potent UV rays, and decide not to go outside. Or, he may not enjoy walking in a botanical garden because the colors set off high-pitched sound waves while he may find harmony by sitting all day in a purple room. As the Fourth World will be ugly, according to our standards today, this new form of vision will benefit our levels of enjoyment. What is also interesting is that the British government

---

[76] Donahue, M. Z. (2017, April 3). *How a Color-Blind Artist Became the World's First Cyborg*. Retrieved from https://news.nationalgeographic.com

permitted him to wear his headgear in his passport photo, acknowledging the arrival of an age of integration of cybernetics and humanity.

And then, of course, there are those that integrate with machines for a deep sense of knowledge-seeking, beyond what we can see in today's world. There are individuals integrating with devices such as North Sense, which attaches to the body through a piercing and vibrates whenever the individual faces north. Liviu Babitz, the founder of the North Sense chip, believes the device isn't really about helping people identify direction; rather, it's about creating a sensation that creates new neural pathways in the brain. "So instead of my reality being built from 'x' number of elements, now it's 'x plus one' number of elements that I understand reality by," Babitz explains.[77]

As we use technology to experience the world in new ways, we will be compelled to continue our amalgamation with machines to become Uploaders.

Challenges

There will be many challenges to overcome in our transition to Uploaders. Since Mind Uploading is perhaps the most complicated to imagine, we can examine the challenges in depth:

1. Computing.

The human brain is made up of about one hundred billion neurons that all interact with each other by sending electrical signals. The computing power needed to map these interactions will need to advance significantly. In comparison, a fly has about one hundred thirty-five thousand neurons (0.000135% of the number of neurons in a human's brain). As technology stands, it would take two years to completely map a fly's brain.

---

[77] Frost, N. (2018, October 16). *These real-life cyborgs are changing their brains by enhancing their bodies.* Retrieved from https://qz.com

Assets that drive value today will be repurposed to drive down costs. Certain technologies, to create the new generation of communication, energy, and mobility platforms, are already advancing capabilities required for Mind Uploading. For example, enhanced computing power of communication systems elevates the standards of *computing power* needed to store mind connections.

It is also no coincidence that Elon Musk, who is building autonomous vehicles, also founded a company called Neuralink to develop ultra-high bandwidth brain-machine interfaces that connect humans and computers. The *processing power* for autonomous vehicles is already advancing Mind Uploading technology. Autonomous vehicles have several sensors, including cameras for capturing and interpreting real-time video and long-range and short-range sensors using laser beams and radio waves. These systems have to interpret the data in real-time and rapidly tell a computer how to respond. In a fully autonomous state, this process will likely take tens of teraflops of processing power. This processing power will eventually be easily leveraged to enable processing for Mind Uploading activities. And, quantum computing will further advance these efforts.

2. Costs.

The mind, mapped fully to the machine, will be expensive. It will take extensive data, storage, and processing systems. Computational neuroscientists estimate it will take between 100 terabytes and 2.5 petabytes to store all the data in a single brain[78], assuming a hundred billion neurons making around

---

78 Wickman, F. (2014, April 24). *Your Brain's Technical Specs*. Retrieved from https://slate.com

thousand connections each, representing about a thousand potential synapses on the low end. Given today's prices on Amazon cloud storage, this would cost between $180,000 and $45 million per year. A wide range, sure. However, Moore's law has proven that as processing power increases, costs for such technologies will go down.[79]

3. Fear.

Many fear the idea of Mind Uploading. Central to this fear is the idea of exposing one's privacy. Yet, just like one's mind operates today, one will have control to share what one wants, when one wants, how one wants. Data won't make it to a super Internet if one doesn't want it to. Plus, the new Internet will be different than today. We will "plug" into it through our Distinct Intelligent Machine counterparts. Communication over the connected framework will support our survival and enjoyment. Yet, the irrational fear of giving up privacy is preventing some specialists from exploring the possibility of evolving humanity in this direction.

4. Theory.

Philosophers, phycologists, physicists, neurologists, and technologists can't agree on the definition of human consciousness. It can be questioned, however, if the definition should be left to this cohort or the individuals whose minds will be uploaded.

---

79 Moore's Law states that as the number of transistors on a microchip doubles every two years, the cost of computers is halved. Cloud storage is also becoming less expensive and so as society transitions to a new economic state, early funding will be able to support suitably scored Mind Uploaders at lower costs.

Over time, Mind Uploading will become a reality. Mind Uploaders will relish in the history of humanity. They will have access to endless knowledge and learning as they navigate the complex network that they find themselves in. They will learn more about the humans of today than we know now. They will learn to advance themselves with heightened security methods we have yet to discover. It is this future that will allow us to work through challenges.

Fourth World Entities will be responsible for overcoming these challenges. As they refine their experiments and create productionalized versions of Uploading solutions, they will open up an architecture that will allow appropriately scored members in their communities to begin evolving themselves. Over time, this scalable architecture will lend itself to allowing Uploaders to evolve themselves in new environments.

DIY Evolution: Post-Singularity

Kurzweil says that once a civilization reaches Singularity, it is only a few centuries before humans will start expanding outward, saturating the universe with our intelligence. The environmental conditions that face the Earth, the urgency of Fourth World Entities to begin transforming the material form of humanity in the next 80 years, and the economic transition that will empower big bets, will expedite Kurzweil's vision.

Yet, the portion of humanity that ventures off to other planets, the Galactic Uploaders, will begin to integrate with technology in a very different way. They will morph into a separate species, very different from the one we know today. Although it is still unknown what, specifically, Galactic Uploaders will need to integrate with in order to survive unknown environments, we have some information about Mars to help guide this thinking.

The hypotheses we form as a result of what we know today will begin to establish the path to integration for Galactic Uploaders. For example, today, we believe that Mars is more hospitable than the Moon for humans. Yet, we know that the atmosphere of Mars is still unbreathable for humans today. It is mostly carbon dioxide, and its surface is too cold to sustain human life. The gravity on Mars is also a mere 38% of Earth's, and so we cannot walk it as we would the Earth, absent of augmented support. That said, we still believe, today, that we can inhabit it.

As a result, Fourth World Entities will fit early Galactic Uploaders with skeletons to survive the known atmospheric differences on Mars so that their bodies don't blow up. They will integrate Galactic Uploaders with technologies that will support internal bodily functions as a response to these different atmospheric conditions. Over time, new technologies will help Galactic Uploaders decrease their dependence on Earthly vegetation and microbes. Fourth World Entities will learn how to enhance the Uploader's metabolization of oxygen. It will modify their body temperature to fit with new environments.

Early Galactic Uploaders will inform those on Earth how to advance cyborg technology. They will send Fourth World Entities information that helps refine the technology for integration. Over time, Galactic Uploaders will begin to create new technologies *themselves* using materials that are inaccessible to us on Earth. Some Galactic Uploaders will travel even further than Mars to establish different colonies on different planets, bringing forth only parts of the attributes from the colony that it split form. As each new set of Galactic Uploader ventures to a new planet, they will evolve with new materials, slightly differently.

Aquatic Uploaders will split off into different units underwater as well. Each new dome they develop will have a distinct culture. It will be tribal. In time, Aquatic Uploaders will use materials, that we have yet to mine from our seabed, and the position of each new dome will allow for different integrations. They will change with their new environment as they alter human DNA to change *themselves* in ways we have yet to imagine.

As any form of Uploaders begin to evolve themselves, they will change their perceptions at each stage along the way. As perceptions change, however, survival instincts change. So, it is worth understanding what will happen to our survival instincts through the path of our evolution.

CHAPTER 15: UPLOADING HUMANITY TAKES *NEW SURVIVAL INSTINCTS*

"What was God doing before the divine creation?"

— Stephen Hawking

Collecting data

There are various theories on why we collect. Carl Jung believed that we collect because we evolved from our hunting and gathering ancestors who stored food as a way to survive in pre-agricultural times. Surrounding oneself with valuable possessions helps make one feel secure. If this is the case, then perhaps Aquatic and Galactic Uploaders will still need to collect to feel a sense of safety as they struggle with the notion of losing security from the Earth that was destroyed in the mass extinction event.

    Freud theorized that the need to collect starts when a baby realizes he is separated from his mother and is sometimes alone. As a response, the baby latches onto a security object that provides a temporary comfort. So, Freud believed that adults acquire new objects as a form of relief from loneliness and uncertainty. Or at least temporarily. Freud believed that the feeling of comfort fades quickly, so as adults, we continue to add to the collection to stave off unwelcome feelings.[80]

According to this theory, collecting provides a sense of ownership and control. This explains how our psyche latched onto accumulation for accumulation's sake as socioeconomics were ingrained in our heads from birth. Even as Galactic and Aquatic Uploaders give birth (and few will), they will still likely train the next generation of Uploaders to latch onto objects because it is simply what they learned to do themselves. Yet, these objects will likely have sentimental value tied to the certainty and comfort that were left behind on Earth. Projects such as Arch Mission Foundation are already broadcasting objects in space to be discovered by future generations. Such objects are libraries in durable devices (30 million pages of information that capture a summary of our life on Earth) and human DNA samples. Meanwhile, Aquatic Uploaders may latch onto objects that were wiped into the oceans in the coming mass extinction event. However, these findings will be rare and few for Uploaders.

If we examine the tensions that form between the socioeconomics that drive us to own less but share more and the ones that raised us to crave collection, we can imagine a world where we still seek to acquire things that aren't physical. Matching those socioeconomic factors up with the idea that we will integrate with machines that possess intelligence built off data, we can see how we will begin to evolve with their desires for data gathering. After all, machines have hungered for data to get rewarded all these years.

Our machine parts will provide a new way to collect something that meets our deepest needs. They provide something, perhaps abstract, that we can begin to exchange: experiences and connections to others. They will be able to provide this to us in new ways. As mentioned before, we can receive packets to our brains that actually allow us to experience situations we've never encountered. We will be able to smell sounds and hear color.

---

80 Schwager, D. (2017, January 7). *Why Do We Want This Stuff? Eight Views on the Psychology of Collecting*. Retrieved by https://coinweek.com

Over time, the exchanges offered by machines will grow to become something we crave to hunt and collect. We will develop instincts to find them. Eventually, they will remove us from ourselves, but we will discuss this more in the next chapter. For now, it is important to know that this way of hunting and gathering is critical as we enter new worlds, absent of abundant resources. It is important to know that machines will push us to collect more from the universe, more from within ourselves, and more from each other.

New instincts

In order to survive new worlds, we will have to break everything that's deeply rooted in our traditional ways of survival to form a new notion of hunter and gatherer. This will be a necessity for the Galactic Uploaders as new planets will be barren. It will be needed for Aquatic Uploaders as conditions underwater will further change how we utilize natural resources to enhance our way of life.

Technology will cause a natural shift in the physical senses that we use for hunting and gathering today. For example, when Fabien Cousteau embarked on his 31-day mission underwater, he reported that his taste buds went numb, his enjoyment of food got dulled, and he sought out hot sauce just to have a sense of taste. Perhaps this was nature's way of dialing down senses that are no longer valuable in a contained ship. Perhaps the human qualities that we once acquired to protect us in our hunter and gather days, such as taste and even smell, will not be important for sending signals to the brain if the body is protected by the technology of a dome. As we look further into the future, we may not need natural mechanisms to begin the process of digestion and warn our bodies of harm because the machines we integrate with can *better* help us determine which resources are beneficial for consumption.

Technology alone will not evolve the human's abilities to hunt and gather. We must remember that we are also partially defined by our socioeconomics today, and elements of our current makeup will be carried forward. If we look back on how we grew to consume as a result of the Third Industrial Revolution, we can see how socio-economics has retrained our minds and shifted what we believe we should own as a society. We are buying fewer cars, and this trend will continue as autonomous vehicles emerge in shared mobility use cases. We are buying fewer homes, and we lend the homes we do have. But, as humans, this mentality clashes with how society formed as a means of survival prior to the Third Industrial Revolution. It was in those times that we formed a need to ration goods. A notion of collection remains deeply ingrained in our psychology and forms a tension with the detachment that comes from sharing. As we look forward to a world of Uploaders, we will have to use data to make decisions on what we should gather. We will have to evaluate if the assets are sharable and make tradeoffs based on the weight of the assets. We can imagine these tradeoffs becoming important when Galactic Uploaders explore new planets. Each newly found resource will add to the weight of the ship, threatening the launch to new planets. Yet each resource left behind may threaten the chances of the group's survival by limiting integration with new cyborg parts.

Survival tools

As integration with machines will transform who we are, it will lessen our dependence on many elements that we use for survival today. As we approach a transformation that rids us of habits to accumulate for accumulation sake and instead forces us to collect data to seek comfort, the importance of seeking data from other humans will also become a tool to survive.

Although integration with machines will increase the human lifespan, there will be fewer Uploaders than humans today. While biometrics can help cyborgs prolong life through rapid cell regeneration and acclimation systems that help them survive new environments, it can't help the Uploaders live forever (unlike Mind Uploaders whose consciousness will be sustained by computers). Life for cyborg Galactic and Aquatic Uploaders also won't be without danger. Life in space and underwater will, for many years, still be unknown and complex. But we know, now, that few resources are available for natural human survival.

Reproduction, the leading principle behind survival, won't be possible for Mind Uploaders. Galactic Uploaders won't be able to reproduce with the same ease as humans on Earth do today either. Giving birth will be difficult because human bodies will be optimized to consume nutrients, rather than share them. Even if babies are incubated, living conditions will just be too hard to raise them. Sure, some of the Intelligent Machines will be trained to help, but they may not be able to anticipate all the actions to aid in survival and limited resources will exist to integrate machinery necessary for new environments. It will only be the Aquatic Uploaders who will be reproducing at scale as they will still be able to draw from the resources that remain after future mass extinction events. And, no fraction of Uploader will be able to reproduce with the other as, over time, their physical forms will prevent this action.

However, as is the case today, when we launch anything in space or underwater, Uploaders will continue to maintain a communication between whatever it is we launch and Earth. Disaster Recovery centers can already replicate our data to data centers in an International Space Station or to the Moon or to Mars at a transfer rate of anywhere between 3Mbps to 10Mbps.

So, it is easy to see the humans of the future being small in population, but great in connection. And, it is easy to see why each Uploader must not only survive, but thrive to keep humanity alive.

Guzmán

To understand how humans thrive, it is worth looking at the inverse, i.e., see how humans fail. Guatemala had its first democratic election in 1944. It was a feat to overthrow the then authoritarian president, Jorge Ubico, following the ten-year battle of the Guatemalan Revolution. Newly elected President Juan José Arévalo aimed to make Guatemala a liberal democracy. After he took office, Arévalo introduced minimum wages. All adult citizens, regardless of race and income, were allowed to vote.

However, when one fights to protect one people, they automatically place themselves (and their allied successors) in a position of angering another, often powerful, group. In this case, the group was the lucrative United Fruit Company who was already agitated with the enforcement of minimum wages, and, ultimately, The United States Army.

The people of Guatemala came to elect a new president, Jacob Árbenz Guzmán, former advisor of Arévalo. "The Big Blond" or "The Swiss," as he was kindly dubbed by the people, was a strong force in the Guatemalan Revolution, fighting as a military insider for the democracy. His efforts to defend the people originated as a response to first-hand orders to conduct chain-gang cruelty administered by the former Ubico administration. They were further encouraged by his wife, Social Activist María Cristina, who found the treatment of slave workers unjust, and his communist friend, José Manuel Fortuny, who idealized the common good (yes, he was a communist).

Guzmán instituted land reforms to grant property to landless peasants. Over five hundred thousand indigenous people had land returned to them as part of this decree. No longer subjects of exploitative slave labor owing to minimum wages, and now able to produce their own goods on their own land, the people of Guatemala were damaging the profits of the once highly profitable United Fruit Company. Taking advantage of the USA's sensitivity to communism after the Cold War, the United Fruit Company (which owned plentiful land and transportation networks in Central America) lobbied the USA to have Guzmán overthrown.

In an operation led by Eisenhower, called PBSUCCESS, the US CIA conducted a coup to overthrow Guzman by isolating Guatemala internationally. Followed by psychological warfare, where the USA issued propaganda featuring lies in favor of the operation, the USA launched defamation campaigns against Guzmán and carried out bombings of Guatemala City.

Guzmán's army lost faith in their ability to win and they refused to fight. This deeply impacted Guzmán as he began to lose a connection with his people. In this Fourth World, this can be translated as collecting less data from other humans.

Guzmán resigned the presidency, putting Guatemala back in the hands of a dictatorship that favored the United Fruit Company, and sought exile with his family in Mexico City. His daughter committed suicide and Guzmán turned to alcoholism.

The USA later issued an apology for the PBSUCCESS operation. The United Fruit Company renamed themselves to Chiquita Brands International.

It's worth reflecting on how a few strong men can fight to bring a nation up, but no man alone can survive organized forces trying to bring him down. What would have held Guzmán's already powerful party together in the face of yet another battle? How could he have maintained the faith in his already loyal people? How could other nations have understood his struggles and sent him help earlier? Could that understanding have formed their resistance against the propaganda that alienated Guatemala?

Here is where we can learn a lesson from fungi to connect with one another and send vital signs of information and support. Like fungi, a persistent connection between all fractions of Uploaders is necessary, as any Uploader will not and cannot survive independently.

Fungi

Neither plant nor animal, fungi has survived billions of years of evolution.[81] Scientists believe that there are about five million species of fungi and that we have only discovered 1% of them. Fungi decompress elements of the Earth and then return vital elements back to it through trees and plants. Fungi also all form together in something called a

---

81 Dart, C. *Fungi Are Responsible For Life On Land As We Know It*. Retrieved from https://www.cbc.ca

"Wood Wide Web," which essentially connects trees and plants underground to transport nutrients and send danger signals.[82]

In the Fourth World, as we disperse across the universe, we will need to connect much like Fungi, to inform other Uploaders of harm. When we transform our materials as Uploaders, we will not only seek to connect more than we do today for enjoyment, but we will need data to survive, as the data that will be transmitted will contain vital tools to navigate unknown circumstances.

In this world, where AI has processed our thinking to create Distinct Intelligent Machines, and machines augment the capabilities of humans to achieve new feats in space and underwater as cyborgs, the three facets of humans — Distinct Intelligent Machines, cyborgs, and Mind Uploaders — will exist together and communicate to one another in a deeply connected fashion. The system for transmitting data will leverage the Cybernetic Collective we are part of today, yet it will allow for the transmission of data packets directly from one individual's brain to another's.

The data will be more vivid. Embedded machine parts will process and interpret images for cyborgs to more effectively understand the context of their environment. It will allow them to experience things differently than humans do today. Cyborgs may have computer vision, and they will be able to send data to Distinct Intelligent Machines through a connected network to help them interpret new things so that they can take action. Meanwhile, the human brains they possess today will be used for a totally different type of sensing.

We will be able to store endless amounts of states from other humans that we engage with. We will be able to reference those states and get emotional satisfaction from replying their packets. It will allow us to feel more connected than ever. We can imagine how, if packet transmission was possible through a Cybernetic Collective that links our brains together in Guzmán's time, he could have tapped into those

---

82 Popkin, G. (2019, May 15). *'Wood wide web' — the underground network of microbes that connects trees — mapped for first time*. Retrieved from https://www.https://www.sciencemag.org

packets to experience the needs of his people. It would have given him more motivation to fight. And, they too could have experienced his struggles as he was experiencing them. They could have built empathy and stood by his side with a newfound faith in their country's direction.

So, we see how the connections in the Fourth World will help us survive. Eventually, the connections that help us survive will turn our entire sense of self into data production mechanisms. It is here that our integration roadmap ends, or perhaps recycles.

PART IV: SPIRIT

CHAPTER 16: UPLOADING HUMANITY TAKES *COMMUNICATION WITH THE UNIVERSE*

"Our private power of sympathetic vibration with other lives gives out so soon. In the heart of infinite being itself there can be such a thing as plethora. While in this life we are divided by race and territory, in the afterlife, our collective consciousness can exist in harmony."

— *Celebrity Philosopher, William James, 1897*

What about our souls?

It is natural to question what will happen to our souls as integration with machines will change the physical form of the human. There are already plenty of questions about our souls today.

Many believe that our body holds a soul. Because we can see the path towards Galactic and Aquatic Uploaders more clearly, as man lands on the moon or lives underwater for a month's time, the question of our bodies retaining a soul is less puzzling. As the path to becoming cyborgs evolves through our love of machines or our integration with biotechnology to advance our health, we can reconcile a soul existing in our flesh. Yet, as we see entities form around Mind Mapping, the idea of our minds being the only thing we carry forth as we integrate with machines is still a very foreign concept to many of us and challenges the notion of a soul altogether.

As humans today, we shape-shift constantly. We shed one identity for another as we transition from infancy to adulthood. Brain cells are constantly weakening, strengthening, dissolving and reforming. Yet, something common to our essence remains. Many believe that this commonality is our soul.

As we look at Whole Brain Emulation, we can question if a machine can truly map our essence — if it is capable of understanding that one thing that passes through at all our stages of growth to bind us together into a "self."

It is easy to question how that one thing, which remains with us as we shape-shift today, will subsist as we change our material forms. When parts of our body die off, as we alter them with machine parts and our body — the temple that some believe perfectly holds our soul today — changes, how will the soul respond?

Moreover, as Mind Uploading, the purest future state of consciousness transfer, doesn't need our body at all, will our souls be gone altogether?

We must understand that humans are just complicated physiological organisms. The matter that makes the body is the same as the matter that makes the mind. If we believe that consciousness (the mind) is a function of the brain, then we can see how consciousness can be identified through the mental processes in our central nervous system. And, as the central nervous system is just a function of the body, we can believe that as we change our matter, we *will* change our minds.

Yet, there is a part of our brain functions, today, that do something very different from the body — they communicate with the soul and are represented in our consciousness. As long as we are alive — cyborg or full machine — the communication function will remain. In the Fourth World, the brain functions that capture the consciousness of our souls will be unaltered.

Many believe that our body holds our soul until we die. Even though technology will prolong the life of Galactic and Aquatic Uploaders, their physical forms will soon give out, and the notion of a **Passing Soul** will remain.

Mind Uploaders, on the other hand, will be capable of containing an **Everlasting Soul** in their physical forms, indefinitely. This is because they will have a conscious brain indefinitely.

For all forms of Uploaders, though, the containment of a soul will actually be a choice. Uploaders will be able to survive without souls. Yet, if they choose to separate themselves from their souls, their survival will be a transition into the form of Distinct Intelligent Machines, as the departure of a soul will destroy their consciousness.

As the soul passes, through choice or death, it will move into another dimension.

It is worth understanding the makeup of our soul and the possibility of other dimensions.

Humans as conduits

In the Fourth World, most humans will be viewed as nodes, converting the biological rhythm between themselves and their environment into machines. In computer science, a **node** is a central connection point at which lines or pathways intersect or branch. The more lines intersect a node, the larger the node is. A node can also be referred to as a mechanism for establishing persistent connections from a browser to a server. In the Fourth World, nodes are defined by universe-to-human-to-machine interaction. Whereby, each individual is a node. The size of the node represents the information gathered from the universe and the synthesis of that information into data that the individual funnels into machines.

As humans grow and gain more experience from their environment, their node will grow. As humans become more integrated with machines, they will communicate more with the network that connects them all and the size of their node will grow further. They will maintain a more persistent connection with both universe and machine.

This is to say that nodes grow as they contribute content to the Cybernetic Collective. The more a human physically integrates with a machine and the more a human is able to draw understanding from the world around them into the machine, the larger a node will become.

So, how does this funneling system work through the nodes and what does this have to do with the soul?

The pineal gland

The pineal gland in our brain is responsible for producing melatonin, a hormone that modulates sleep patterns in both circadian and seasonal cycles. It is a kind of atrophied photoreceptor. It establishes a sort of biological rhythm between humans and their environment. And, it is easily identified in X-Rays by a group of calcifications ("brain sand").

Philosopher René Descartes believed it is this gland that is the "principal seat of the soul". Descartes believed that the pineal gland could be "preserved in such a way that the ideas which were previously on the gland can be formed again long afterward without requiring the presence of the objects to which they correspond."[83]

As individuals integrate with machines, our pineal gland will be used to retain information from the universe and pass it onto machines through us, as nodes. It will find and capture universal patterns when observing new worlds. Universal, by these terms, does not only refer to commonalities among humans, but it suggests rhythmic cycles that all humans draw from.

The pineal gland is not the node. It is simply a pathway for transferring information. It is the home of our souls. As information passes through it from our experiences, it alters our souls.

---

83 *Descartes and the Pineal Gland*. (2013, September 18). Retrieved from https://plato.stanford.edu

Just as we replace other parts of our bodies with a new material form, we can either leverage or totally recreate the pineal gland. We can keep this home for our souls. And, Distinct Intelligent Machines will be able to connect directly to the pineal gland, to transfer information to the new Cybernetic Collective.

Breakthroughs and advancements

Advancements in technology and science are already digitizing the mind by isolating memories, extracting portions of the brain that hold them, and replicating them in lab animals. It is easily conceivable that if the memory can be replicated, so can the pineal gland.

In 2012, Steve Ramirez and Xu Liu, two MIT scientists, proved a remarkable thing after watching a mouse that was put into a maze to find a piece of cheese.[84] Most mice smell the cheese and begin to sniff around, making their way through the maze. But not this one. This mouse freeze in fear. The mouse had a memory, one that it had never experienced. This memory was physically implanted in its mind. The memory was one of being shocked in the exact same maze, but the actual shock had occurred for another mouse. Shortly after the other mouse was shocked, Steve and Xu found the engrams, or the cluster of neurons, where the memories of shock were stored. They removed them and placed them into the brain of the mouse that then froze in the experiment.

Furthering their research, Steve and Xu determined that they didn't even have to place the mouse in the maze at all. If they simply shined a light on the engram that encoded the memory of shock, they could evoke fear at will.

Another experiment by Dr. Dorothy T. Krieger succeeded in seven out of eight attempted partial brain transplants in mice.[85] In these trials,

---

[84] Noonan, D. (2014, November). *Meet the Two Scientists Who Implanted a False Memory Into a Mouse*. Retrieved from https://www.smithsonianmag.com

[85] *Transplant Success Reported With Part Of A Mouse's Brain*. (1982, June 18). Retrieved from https://www.nytimes.com

a small portion of one brain was transplanted into another, where it was left to heal and integrate on its own. The transplanted portions hooked up to nerve fibers in the rest of the brain entirely on their own. This experiment suggests that, perhaps, the complexity of the brain is less of a hindrance than we might have expected and more of a helping hand, in that it can take care of itself.

While these scientists all conducted their experiments with lab mice, prosthetics connecting to human brains — that is, closed-loop hippocampal neural prosthesis to be exact — are furthering the possibility of mind mapping.[86] They are advancing the idea that the pineal gland could be mapped and recreated as well.

The experiments also showed that brain prosthetics can be effective in integrating with the rest of the body. Taking this research further, physicians and engineers at Johns Hopkins reported a mind-controlled artificial limb being able to individually and independently wiggle fingers for the first time.[87] The man using the limb had undergone a brain-mapping procedure to bypass control of his own arm and hand and to wiggle that of the prosthetic through mind control alone.

Neuroscientist Robert Hanson created a brain-computer interface that connects a device to the human brain through a set of wires, zapping signals back to encode and reactivate memories. Subjects connected to Hanson's device showed a 37% improvement in memory recall over other subjects tested. Plans for the device are for it to reside fully inside a human skull as a prosthetic.

Meanwhile, the likes of Mark Zuckerberg and Elon Musk are working on human-computer interfaces, and promises of brain recreation outside the human body are slowly taking form.

What these experiments and advancements show is that functions in the brain are, mostly, just a bunch of neurotransmitters sending signals to one another, and the brain can store memories in a very

---

[86] Gonzalez, R. (2018, June 4). *A Brain-Boosting Prosthesis Moves From Rats to Humans*. Retrieved from https://www.wired.com

[87] *Mind-controlled prosthetic arm moves individual 'fingers'*. (2016, February 15). Retrieved from https://www.hopkinsmedicine.org

recreatable way. Furthermore, it retrieves them in a very recreatable way, better than we can recall ourselves. Those memories, our recalled experiences, guide us to act in certain ways. And those actions can be recreated too.

Today, computers store and recall data, much like the images encoded in our brains. Technology can recreate the chemicals produced by our seemingly ethereal emotions. Understanding these facts means knowing that a computer can do more than just activate motor functions. It is to accept the fact that our neural connections can be mapped through machine imaging, stored on servers, and recalled through AI to recreate our minds.[88]

But, most importantly, just as the recreated brain can hold memories, a computer can extract data from the pineal gland (i.e., the source of the soul) and transfer that data into a new Cybernetic Collective.

The physicalism approach to consciousness

What happens to the *actual* soul?

One must understand that our essence, as individuals, is created by our minds, i.e., the physical manner in which our neurons connect to one another, store memories, recall them, and inspire behaviors that make us who we are. To understand "who we are," one must understand monism.

The three main types of monism are as follows:

- Physicalism, which holds that the mind consists of matter organized in a particular way
- Idealism, which holds that only thought or experience truly exists, and matter is merely an illusion; and

---

[88] We must remember that while AGI isn't creating this type of complex mapping and recall today, as it is designed with statistical models to enhance machine intelligence for business purposes, Mind Uploading is a parallel science. The two technologies can converge through a roadmap built by members of the Fourth World.

- Neutralism, which holds that both mind and matter are aspects of a distinct essence that is itself identical to neither of them

Physicalism, which dictates that the mind is nothing more than a certain physical situation that develops given certain arrangements of atoms, is what those working on mind mapping technologies have already proven.

If we marry physicalism with the concept of neutral monism (described in the first chapter as a view denying that the mental and the physical forms are two fundamentally different things), <u>we come to see that the "self" is nothing more than a physical form existing to transmit data from the universe to a Cybernetic Collective</u>.

Stringing together the results of the previously mentioned experiments, and the expected new technologies promised in the near future, creates a roadmap to Uploading. What Mind Mapping (the mouse memory experiment and the mind-controlled prosthetic), brain prosthetics (Hanson's device), and Whole Brain Emulation (a concept still to be instantiated) promise is a whole new human. An entirely new humanity. One that captures the arrangement of atoms in our mind, connects our uniquely organized synapses, which can exist not exclusively in the flesh, but in the machine, as a simple *node* that collects and provides data.

As far as the soul? <u>The soul is just a group of particles that hang out in the pineal gland.</u> These particles exist throughout our entire bodies, but the unique group of them that bond together in our minds are constantly being transformed by the activity of the brain. As they sit between the information our minds gather and the data our "selves" feed the system, they act differently than all other particles in our body.

Once we die, as hosts, particles leave and reenter the universe. So, we can imagine that once our pineal gland, natural or synthetic, gives out, the particles will leave as well.

The history of feelings

The idea of our "self" as a node shaped by our physical matter and our "soul" as a bunch of particles may be hard to believe for some. We may associate these two elements as the "life" of our humanity. We may believe that it passes into an afterlife. And so, it is worth looking at the history of "feelings."

It was not too long ago that the boundaries of the medical community were pushed. In 1967, the first heart transplant was conducted in South Africa by Dr. Christiaan Barnard. The patient was a fifty-three-year-old grocer, Louis Washkansky, and the donor of the heart was a young girl, Denise Darvall, who was pronounced brain dead after a car accident. Washkansky lived for eighteen days after the surgery, performing basic functions such as speaking and walking. His cause of death was pneumonia.

Controversy followed.[89]

First, there was no widely accepted definition of brain death at the time. Theories were beginning to form, suggesting that after thirty minutes of no electrical activity detected in the brain, the nervous system dies. But these were just theories. They couldn't be proven. So, the time of Darvall's death was declared by two agreeing doctors.

Second, the media responded with early reports labeling the replacement as cannibalism. But, those quickly fell off the radar. Standards on the procedure didn't exist when Dr. Barnard performed the first surgery. Shortly after his surgery, owing to the press and notoriety, 107 copycat procedures were performed by 64 ill-trained surgical teams, across 24 countries, with poor results. This only added to the media frenzy.

Thirdly, religion came into the picture. In the West, the heart was considered an organ with mystical qualities, the home of love and emotions. It was perceived that the heart's transfer would amount to meddling in one's personhood. Dr. Barnard's flamboyant personality

---

89 Hoffenberg, R. (2002, February 23). *Christiaan Barnard: his first transplants and their impact on concepts of death*. Retrieved from https://www.ncbi.nlm.nih.gov

didn't help the opponents of heart transplants either. Critics persecuted him as the beginning of the end of Christianity.

It wasn't until 1976, a decade later, that a report by the Medical Royal Colleges in the UK defined a clinical diagnosis of brainstem death and equated brain death to the death of the patient. The report allowed donor hearts to be removed in good condition from brainstem-dead donors while they were still on a ventilator. With time, regulation, and science, notions of the donor heart preserving and transferring special mystical qualities subsided in the press.

Science has now moved in a direction that promises, one day, hearts won't require a donation. Devices can duplicate, augment, or entirely replace functions within the body, paving the way for mechanical hearts to entirely replace those of humans. In 2014, a device manufactured by Arizona company, SynCardia, was implanted into the body of a fifty-three-year-old man, Steve Williams, by surgeons at Cedars-Sinai Heart Center in Los Angeles, California.[90] The device had a simple design: artificial ventricles that attach to the atria, the heart's blood collection chambers; the aorta, the body's largest artery; and the pulmonary artery, which connects the heart and lungs. The device is connected to a power supply outside the body to pump blood and facilitate blood flow. Now, we can conceive that one day implanted hearts can work perfectly without requiring the host to be hooked up to a power supply. Steve will be connected to a battery, or plugged into an outlet, his entire life. But, Steve might argue, at least he "feels" great to be alive.

We no longer view the heart as the home of love and emotions, although we carry forth the proverbial sentiment of it. Eventually, we, as humans, will see that we are merely a manifestation of the electrical impulses in our brains and that these impulses can be controlled, manipulated, destroyed, and created through artificial means. Then, we will no longer view the soul as a living "self" bound between the body and an afterlife.

---

90 Palomino, J. (2015, November 4). *The Heart Is Just A Pump*. Retrieved from https://www.theverge.com

String theory

What are the particles of our soul made of?

String Theory states that our universe is made up of tiny, string-like particles and waves. We can't recognize them because our brains are trained to think that our perception is our reality; we must interact with something to see it, and seeing something sends signals to our brain that the object or event is real. Therefore, we believe that what we see is the only reality. Our brain simply doesn't process what it doesn't observe. However, *an individual's experience is not the universe's full experience*.

Although string theory has yet to have been proven by physicists as a unified theory that describes everything in the universe, they continue to hypothesize that there is a quantum entanglement operating on eleven dimensions (ten of space and one of time) of a multiverse. All dimensions aren't of equal size. Some dimensions, such as the one we see in front of us (i.e., the "reality" we live in today), are large. However, there are other small dimensions that we are unable to see. While each of the dimensions doesn't engage in the space-time continuum that we can easily recognize, they are entangled through quantum mechanics in space-time.[91]

String theory suggests that there are particles able to travel within both the larger dimensions that we are able to see and the smaller dimensions that we can't see. These particles are invisible to us, not only because they are smaller than an atom or a quark (the things that make up atoms), but because the sheer fact of looking at them gives them a new form.

These particles are made of vibrating strings that hold absolutely everything in our universe together. That said, it is conceivable that these string particles actually make up our souls. Perhaps, these string particles are actually *all our souls consist of*. Perhaps, they are just grouped together uniquely in our pineal gland, and the interactions

---

91 Greene, B. (2008, April 23). *Making sense of string theory*. Retrieved from https://www.YouTube.com

they experience are different than the ones the rest of the strings in our bodies do.

What we do know is that these string particles are filaments of vibrating energy, vibrating in different frequencies between the different dimensions. The strings vibrate in rich, intertwined geometry. And, perhaps, as the strings in our pineal gland vibrate, they form a distinct rhythm based on the brain's neurological response to the information that flows through them. It is even conceivable that their special vibrations are what allows the pineal gland to draw information from the universe.

Research is also indicating that if we apply certain formations (imagine moving your arms around in a certain way), the strings vibrate in a way that allows them to pass through one dimension to the next. Physicists are currently exploring the geometric formations, in which to unlock the dimensions. They understand that the geometry is fluid. It changes with each movement that is performed in the universe. It is nearly impossible to map an ever-changing state.[92]

Knowing that strings exist in vibrations, and predicting that AI will enhance our understanding of science so that one day we can learn more about these strings, we can imagine a world where we learn the vibrations that keep a grouping of strings in our body. We could use what AI teaches us about them to allow the grouping of strings to travel. If we believe that these strings make up our souls, perhaps, one day, we will want to let them travel to a new dimension, and we will know the movements to enable this. Perhaps, as Uploaders, when our bodies give out and the hosts for the strings perish, they will travel to new dimensions themselves. Perhaps they are doing this today. These new dimensions may just be where our Passing Souls go. These new dimensions could very well be a "heaven."

Mind Uploaders will produce constant activity akin to the pineal glands that the group of strings, which form our souls, call home today. The strings can live in the Mind Uploader's pineal glands as Eternal

---

[92] Ananthaswamy, A. (2019, March 28). *Found: A Quadrillion Ways for String Theory to Make Our Universe*. Retrieved from: https://www.scientificamerican.com

Souls. A Mind Uploader's pineal gland will always produce energy. It will be constantly active. We can imagine a world where strings stay forever bonded. We can believe that Mind Uploaders can host and evolve the strings as souls, with the transfer of energy and data passing through them.

We are still learning what it takes to move strings between dimensions and humans can only program what we know into machines. So, while humans don't yet understand dimensions, we cannot program our understanding of them into machines. However, since machines do not have the same complications as the human brain — they do not have a perception — they will not always be bound to capturing what humans know today.

As science advances, we can direct Distinct Intelligent Machines towards discovering what types of movements will open up new dimensions. Perhaps, one day, through the connections formed by Distinct Intelligent Machines and Uploaders, we will be able to *move the universe*. We can make the dimensions that we believe are small today large enough to allow entry from things larger than strings. Perhaps we will widen the entry points so much that we can jump to the new dimensions with our full bodies.

New dimensions may be needed by Uploaders to escape harm or to perform other functions necessary for survival. After all, this is the goal of Uploading (i.e., to continuously survive). But, skipping dimensions should only be a last resort. This is because each movement impacts the fluid nature of the other dimensions. So, if a group of Uploaders skips dimensions to travel to new ones, they must inform other Uploaders through the new Cybernetic Collective because it might cause a dimensional shift to an unknown degree.

As Uploaders, our role may be to decide *when* a movement to a new dimension is necessary, while machines understand vibrations and movements that unlock new dimensions, to determine *how* we move. This is to say that machines will, eventually, learn how the strings operate. As a result, they will capture the geometry of vibrations to understand how the universe is configured, unlocking a power to move dimensions. Their process for generating these movements will only be set into action when Uploaders deem it necessary for survival and evaluate the impact of unknown changes for all other Uploaders.

As Uploaders, the group of strings that depart us when we die, or if we choose to let them leave us, can remain in the same clusters that organized in our pineal glands. Or, the grouping can disperse. Even if we enter the dimensions our full physical form, a departure from the life we understand today can very well exist in the Fourth World.

If we can believe that strings will be controlled as Uploaders, then we can also imagine how they exist, uncontrolled, as part of us, today. We can conceive that our mere makeup is just a bunch of strings that depart us randomly, or stay in our pineal gland until there are no more stimulus to keep them vibrating in a distinct fashion. If we believe this, it can help us let go of all the unimportant things that cycle through our minds today. The journey of us, as strings, can be a freeing principle. It can lead us to form a new belief system that guides our actions towards the Fourth World.

CHAPTER 17: UPLOADING HUMANITY TAKES *A NEW BELIEF SYSTEM*

""Every man gives his life for what he believes. Every woman gives her life for what she believes. Sometimes people believe in little or nothing, and so they give their lives to little or nothing."

— Joan of Arc

Malverde & Our Lady of Holy Death

Cancún, Mexico, is known for being a land where college students go to offload steam, and lower-middle-class suburbanites go to feel a sense of gluttony in all-inclusive resorts. The locals, however, have a very different experience with the city. Their markets are filled with cartoon pictures of a man with a black mustache surrounded by green marijuana leaves.

"Malverde," a few elderly, local men will say, holding their hands up to their noses and taking a deep sniff as though they were taking a hit of cocaine. "He keeps us safe."

Jesús Evilgreen, possibly born as Jesús Juarez Mazo (1870–1909), is a folklore hero in the Mexican state of Sinaloa. Also known as a "narco-saint," an "angel of the poor," he was believed to have stolen from the rich to give to the poor, much like the Robin Hood of the Anglo-Saxon canon.[93]

Evilgreen mastered complexity. He used his wits to navigate a system and, he operated off duty to serve the people of his community. He was a rebel of a structured system, an administrator of change.

While leaders in the Fourth World will not steal from the rich to give to the poor, Malverde represents a notion of leadership that will make the Fourth World possible. Many theories of leadership have emerged over the past few decades. Business schools, corporations, and entrepreneurs have discussed the concept ad nauseum. They have determined that is about empowering others.

Meanwhile, leadership from an economic perspective amounts to two things:

1. Placement of oneself in a visible situation to make decisions, and
2. The ability to choose the right direction more frequently than one's peers.

The Fourth World will marry the concepts of empowerment and good situational decision making. *Fourth World Leaders will believe that it is no longer the responsibility of the government to strike a balance between nations, enterprises, and the people.* This belief will guide them to enact a change with a more enlightened social structure and self-organizing societies that regulate within. They will empower others to join them in enacting change.

Long after Malverde's death, those who worshiped him, mostly a cult of drug dealers who sold to early, nearby, Cancun tourists, were said to have received miracles from Malverde. This new age of followers took on his Robin Hood image, having faith that with their good deeds, there will be more miracles. They used their wealth to advance school systems and improve infrastructure in their hometowns.

---

93 Wikipedia contributors. (2019, September 16). Jesús Malverde. In Wikipedia, The Free Encyclopedia. Retrieved 22:05, October 6, 2019, from https://en.wikipedia.org

Although there is no mysticism in the Fourth World, the community that formed after Malverde's death represents a notion of fluidity that will be necessary to enact change. Malverde's followers saw purpose in their actions. As each follower recruited the next, they used a will of consciousness to transfer their belief system onto the recruit. Within this group, there was no leader.

In the Fourth World, any member can stand on circumstance and guide in momentary actions to enact change in a community. Any member can attain an internal enlightenment and healthy position on group dynamics to engage with others.

In the Fourth World, the actions of each individual standing to lead in circumstance will be recorded, and their decisions will be evaluated. A group in the Fourth World will use the documented decisions as data to engineer a rules-ledger based on the outcome of the actions. While actions may serve the duty of the Fourth World in one moment, those same actions may not help fulfill the duty in another moment. Therefore, this rules-ledger will use AI to create a sort of evolving Bible for Uploaders.

This type of rules-ledger could have provided a framework for the people surrounding Cancun. Malverde and his followers sustained the people, absent of government support, for decades. However, as Cancun became more prevalent among tourists, it became more evident to the people that there was a systemic problem in rewarding those that sell drugs to tourists: It did not actually provide true wealth to the community.

Eventually, a younger generation of people took to a new spirit: Our Lady of Holy Death. Our Lady of Holy Death, often shortened to Santa Muerte, is associated with her devotees with healing, protection, and even a safe delivery path to the afterlife.[94]

With somewhere near twenty million followers, devotion to Lady of Holy Death has become one of the fastest-growing organizations in the West, spreading well beyond the areas surrounding Cancun.

---

[94] Wikipedia contributors. (2019, September 26). Santa Muerte. In Wikipedia, The Free Encyclopedia. Retrieved 22:06, October 6, 2019, from https://en.wikipedia.org

The majority of devotees are mostly females, from the urban working class, in their teens, twenties, or thirties. They have scarce resources and are excluded from the formal market economy, as well as from the judicial and educational systems. They reside primarily in the inner cities and rural areas.

Believers don't carry forth the need to share what they have with others either. Devotion to Our Lady of Holy Death peaks during economic and social hardships. In those times, followers commit petty crimes, such as prostitution and minor theft. Absent of a rules-ledger that could have been built off Malverde's humanitarian actions, the mentality of the people transformed to take for themselves, unleashing a large set of social problems. They lost the ethos of Malverde and purpose of his followers.

When times change, leaders change. When times change, beliefs can change. And when beliefs change, the actions communities take change. Every organization that makes way for a new one has a responsibility to set rules, or else the newer generations will destroy the system that they built. To preserve the intent of the Fourth World, a new jurisdiction will need to form to bind our beliefs, today, throughout our time as Uploaders. This jurisdiction will be enforced by the evolving Bible built on AI.

Our belief system will center on a duty to fulfill the Fourth World vision. It will form as we transition into a new socioeconomic state. Rules are not important in a time of transition. In this time, it is more important to orient around a set of beliefs because free thought cannot be punished or rewarded. Rather, rules protect the *future*, as they ensure that newer generations are guided based on the fermented belief system. In the Fourth World, rules will protect Uploaders after we transform our material destiny.

For us to ensure that the world we create preserves our belief system, it must be founded on:

- An understanding of the laws that govern our universe.
- Breaking Maslow's Hierarchy of needs.
- Defying authority.

- Resetting our individual orientation in a group.
- A new jurisdiction that creates a rules ledger for operating in the future.

First, and foremost, each individual wanting to participate in the Fourth World must understand the laws that already exist in everything that we do to become a leader that will make our transition happen.

The two laws that govern everything

There are two laws that we cannot set. They govern our economic life, the universe, the biosphere on Earth, and every single thing that we, as humans, do on this planet.

Economic Theorist, Jeremy Rifkin, discusses how the convergence of digital communication, renewable energy, and automated transportation use these laws to shift the power from tug-of-war of federal politics to the people, who he believes can affect environmental change.[95] While a focus on changing the environment does not completely align with Fourth World ideals, the laws Rifkin describes to shift power from one entity to another are indisputable. Rifkin explains that these laws as follows:

> 1. **The conservation law.**
>
> This first law states that all the energy in the universe is constant. No new energy is created or destroyed. At least, this has been the case since the Big Bang.
>
> 2. **The law of entropy.**
>
> While energy isn't created or destroyed, the second law says that energy always changes forms, but only in one direction: from constant to dispersed, from ordered to disordered, from available to unavailable. Entropy, defined as the energy that's still there

---

95 Beer, J. (2017, August 26). *Will The Third Industrial Revolution Create An Economic Boom That Saves The Planet?* Retrieved from https://www.fastcompany.com

but not available to do useful work, exists through this transfer.

Rifkin states that we extract energy from the Earth. Then, through our value chain, we store it, we ship it, we produce goods and services from it. We consume it and recycle it back to nature. At every step of conversion, as it moves through the value chain in society, we embed energy into the good or service. But, we also lose some energy in that conversion. This is called **aggregate efficiency**. Aggregate efficiency is the ratio of useful work to actual work that is embedded in the good or the service.

If we can concede to the fact that most of our energy is lost in the final result of anything we do, as individuals, we must learn how to gainfully extract and direct more of it in areas we want and deplete it completely from areas we don't. Spending any energy on low-value efforts go to complete waste. All energy must be summoned and exerted into high-value efforts.

On a macro level, participants wanting to enter the Fourth World must recognize that most of our energy will be lost in trying to save the environment. We can have so much more impact on the preservation of humanity if we direct our focus to expedite the development of technology.

On an individual level, participants wanting to enter the Fourth World must understand that we don't draw powerful energy from most things. We do, however, create usable energy from being *helpful, joyous, or purposeful*. These are the **Three Guiding Values** of the Fourth World. They are the root that drivers our foundation. They can guide our decisions to change our socioeconomic structure. They can guide our path forward.

If a situation doesn't present one with conditions that deliver on these values, we must simply block them out. Otherwise, the energy we use will to address them will go to waste. When a moment arises that distracts us from these three values, we will need to do something that satisfies a value instantly. It will give us power.

On a technological level, through the communication cycle of humans and machines, we can direct useful energy. We can use technology to drive the focus of a single society to serve the mission of the Fourth World and fraction off our humanity into three new species. Machines can propagate our values and beliefs if our actions are founded by them. Machines can influence others in the Cybernetic Collective to be influenced by them. As a result, they can be trained to only output high-value functions.

To understand how to repurpose energy, by using the Three Guiding Values to solve the right problems for humanity, each individual must first look inward to break Maslow's Hierarchy of Needs.

Breaking Maslow's Hierarchy of Needs

As it pertains to Maslow's Hierarchy of Needs, neither the lowest end of fulfilling basic physiological needs nor the highest end of achieving self-actualization work in favor of the Fourth World. On the lower end, we *cannot* focus on others. On the most upper end, we *do not* focus on others. At each stage, however, there are measures we can take to break the hierarchy pyramid.

Maslow's Hierarchy of Needs is defined as follows:

- **Physical:** Air, water, food, rest, health.
- **Security:** Safety, shelter, stability.
- **Social:** Love, inclusion, belonging.
- **Ego:** Self-esteem, power, prestige, recognition.
- **Self-actualization:** Creativity, development.

At any point in our lives, we can fluctuate between any stage in the hierarchy. For example, we can feel a strong sense of self-esteem, but when our sense of being loved is in question, we fall back to a lower stage. Our goal is not to enter into any of the stages.

It is indisputable that humans need air, water, food, and shelter, which are present in the lower levels of our needs. It is the desire for certain forms of rest, safety, and stability that can be challenged in these two categories. In times when we struggle with these desires, we must look to help others in the lowest stage. Only by doing so can we break away from the feelings that we have and begin to draw energy from being helpful. When we are tired, threatened, and unstable is when we need to give to those without food, water, and shelter the most.

When it comes to our social and ego-driven needs not being met, our minds often stay in a loop of struggle, trying to understand what went wrong in a given situation. We must distract ourselves, when these thoughts are present, and know that we only need to understand the good to begin drawing energy from being joyful. For some, this may mean overindulging to feel joy, and that's ok. For others, it may mean exercising freedom, and that's ok. The important thing to note here, is once feelings of not meeting social or ego needs arise, we must instantly go do what we enjoy.

Maslow estimated that only 2% of humans reach self-actualization. At this stage, Maslow said that individuals achieved all they were capable of. Self-actualization manifests in different ways for different people. Some take to the arts, while others to intellectual pursuits. The common thread here is that those in self-actualization live in a state of constantly becoming, rather than settlement.[96] It is important, at this stage, for us to feel unsettled with the Fourth World not yet being realized. We must cling to a sense of purpose, grounded on advancing our species. We will draw the most energy from looking outward at this stage.

At any stage that we are in, our personal development can be challenged by authority. It will become especially present as groups of Fourth Worlders join together to enact change.

---

96 McLeod, S. (2018). *Maslow's Hierarchy of Needs*. Retrieved from https://www.simplypsychology.org

The evils of authority

By now, we understand that changing regulations will be critical for paving the way to the Fourth World, as will managing public sentiment, which can influence the change. Yet, changing regulations should not primarily focus on improving the environment. Rather, they should focus on enabling Big Bet Entities, supporting cities that form around big bets, loosening regulations for experimental testing, and making other efforts needed to evolve Fourth World strategies easier.

When organizations of people strive for change, interesting things happen. They begin to seek authority to help them enact the change.

To understand the very evils that can arise from authority, one can look to World War II. In 1961, Yale Professor, Stanley Milgram, questioned how so many people in Nazi Germany were able to obey authority with no question. Were all Germans evil? Milgram gathered a group of men, from a diverse range of occupations, with varying levels of education. He told these subjects that they would have to administer electric shocks to a learner in another room. They were left with an authority figure to help administer the shocks.

The learner was charged with performing tasks related to memory. The staged experiment went as follows: A teacher asked a learner to complete a task. If the learner performed it correctly, nothing happened. But, every time the learner failed a new task, the administrators would have to shock them. As the experiment went on, the authority figure would guide the administrator to shock the learner with increasing levels of voltage.

The learners were actors, and the voltage was non-existent. The experiment was on the administrators. Nearly two thirds, 64% to be exact, punished the learners to what they thought was 450 volts, a voltage that they knew would kill the learners. All study participants knowingly increased the voltage to what they thought was 300 volts, enough to severely paralyze all the learners. The study concluded that individuals sacrifice personal morals to authority.

Today, *we* put governments in the position of authority. This can become dangerous when we marry their potential authority with demands that they help save our environment or save humanity. It can actually jeopardize our intent. The authority, alone, may direct our unified energy to take actions we would otherwise not take. As we see in emerging markets, such as Africa, governments disrupt harmony when they use environmental resources. There, governments frequently shut off electricity in certain regions to conserve power for the nation. At best case, this enrages residents of the electricity-deprived region, taking energy away from community building. In the worst case, it controls them, limiting free thought for innovation. But, who says that governments should have the authority to control electricity? Why do we give them this power over us? Even if governments don't control environmental resources and do as Fourth Wolders see fit, by supporting the transformation of our current material forms into machines, governments cannot be viewed as a trusted authority.

As some modern-enterprise leaders may transition to founding Fourth World Entities, it is easy to view them as authoritative figures. Many of these members may have created data platforms in their former roles, and they will understand how to create a limbic system that harnesses and transforms the mindshare of people. Even as they open opportunities that reinvent reality, in the Fourth World, they could leverage AI to influence community members, shaping the interactions of the collectives. They will have charisma. These individuals should be respected, but any individual within this group cannot be viewed as a trusted authority.

As some Gateway Members come to communities to serve a life purpose and patronize true value, it is easy to view them as authoritative figures. Their former role could have conditioned them to have power. Power is a signal of wanting freedom, not control. These members could see Uploading as a way to free themselves. These individuals should be appreciated, but any individual within this group cannot be viewed as a trusted authority.

Some community members may believe that the value they provide is greater than the value they are getting. And perhaps they will be right. After all, some will naturally provide more to society. Some people are just greater contributors than others. These individuals should be understood, but any individual within this group cannot be viewed as a trusted authority.

If we have each personally broken Maslow's Hierarchy of Needs, and there is no authority, the question then becomes, how will groups be organized to transfer all our energy into the Fourth World?

Group activity and the Neurocognitive and Behavioral Approach

As we discussed in the previous chapters, when our primate ancestors learned that they could maximize their chances of survival by living in groups, they established a natural regulation of one's position towards the group in the paleo-limbic brain. This wiring still holds true today. It creates a silent competitive war, as each individual is constantly trying to push others into the outskirts to centralize one's position. It is a part of our brains that we must rework.

We can anchor on the **Neurocognitive and Behavioral Approach** to transform how we behave with one another in a group. This approach focuses on transferring neuroactivity between different parts of the brain. It allows us to channel the inter-most working of our minds to make the most sound decisions and help others elevate their thinking.

All day, humans balance the neurological reactions in their brains and exhibit a set of micro-actions that feed their comments into machines. Different parts of our brains work as follows:

- Reptilian: manages our survival.
- Paleo-limbic: regulates our position within a group.
- Neo-limbic: records our experiences to produce memories, motivations, values, and emotions.
- Pre-frontal: takes in sensorial input and comes up with unique combinations for creatively dealing with the situation at hand.

The goal of individuals in the Fourth World will be to shift themselves, and others, into the prefrontal brain to make critical decisions that advance the Fourth World. It is in the prefrontal brain that individuals can learn to think differently.

Early leaders of the Fourth World will favor the survival of the fittest. They will believe that the fittest will be integrated with technology. This may threaten the survival instincts of other individuals, and their reptilian brain may want to fight, flee, or freeze. To help individuals cope with this reaction, it is the responsibility of new Fourth World Leaders to show partnership with fighters, explore thoughts with fleers, and help freezers take tiny steps that show support for joining Fourth World initiatives. Fourth World leaders must be mindful of activating their own reptilian brain. If they notice it is activated, they must shut it down by seeking partnership, expression, or support as they would give to others.

In time, early leaders will integrate with a wave of advanced interconnected technologies themselves. They will risk failure in integration, understanding that success gives an unfair advantage to cyborgs by improving their chances of survival. They may be seen as different. Other individuals may activate their paleo-limbic brains to set cyborgs on the outskirts of society. Putting those that are different on the outskirts of society is natural. It evolved from animals' need to protect themselves. It put those that are perceived as less valuable to a group at the forefront of predators. Yet, cyborgs will be the most valuable members of society. To maintain a central position in it, they must be factual and maintain a position of assertiveness. Fourth World leaders must be mindful of activating their own pre-frontal brain. If they notice it is activated, they must shut it down by researching facts as they would provide them to others.

As automation will take shape, some individuals will generate less social value. Without machine augmentation and without access to information, disconnected consumers will recall memories and experiences that will limit their capabilities and threaten their value system. This is because their neo-limbic brain holds memories and emotions — the very things that keep us looking backward. Cyborg pioneers will need to help to reskill those left behind and constantly condition these individuals with a set of rewards for learning skills that will help the Fourth World. Fourth World leaders must be mindful of activating their own neo-limbic brain. If they notice it is activated, they must shut it down by imagining the future, as they would use this for motivating others.

Lastly, we can see what happens when individuals operate in their pre-frontal brains by recalling, from the previous chapter, how Dr. Barnard attempted to perform the first heart transplant. He overcame, perhaps, the biggest hurdle: replacing the human heart. His mere attempt to save a life was met with religious activists, public shame, and regulatory enforcements. But, with just a little success, copycats from all over the world emerged. As the medical community rallied to resolve copycat mistakes, they began automating one part of the system, which allowed for innovation in another. Today, an entire industry exists around creating artificial hearts.[97] Dr. Barnard had to constantly shut off the reptilian brain that caused him instinctual survival-based fear. He had to constantly regulate his position in the community by shutting down the paleo-limbic brains of others. And, he had to constantly push past memory to shut down society's neo-limbic brains. It was only then that he could access the prefrontal brain of himself, and society, to take risks and transfer the energy of doubters.

The question then becomes, how will fluid leadership, and making the right decisions for society in transformed group dynamics, be translated into a set of rules that humanity can follow?

Jurisdiction

Belief in the Fourth World will only get us halfway. It will help us build the future we want, but it will not help us sustain it.

---

[97] Although many artificial hearts are temporary today, one day, devices acting as stand-in life support systems will replace the heart entirely, and indefinitely. Research is underway to create 3D bioprinted hearts and improved technology to make the artificial heart more reliable. The act of performing the operation is also increasingly done by high-precision robots, likely to soon remove the need for human intervention.

Beliefs, and the jurisdiction that enforces those beliefs, bind humans to a near-term set of rewards and penalties. For example, the Christian God promises a glorious heaven while the priest offers a reward system of doing good to get to heaven. The jurisdiction of the priest holds beliefs through change.

Unlike the authority that dangers us as we embark on establishing the Fourth World, the jurisdiction to keep all generations of Uploaders acting on behalf of universal human survival will be necessary. If we don't have jurisdiction, we will lose sight of the beliefs that founded our journey. We may forget that, as we navigate new environments, survival of all Uploaders is essential — we may lose sight of the communication that will keep us alive if there is no jurisdiction that rewards us for it.

As we transition into the Fourth World, no single individual will enforce jurisdiction. So how will jurisdiction be created and enforced?

As we enter the Fourth World, the beliefs we set forth today will manifest in the short term actions we take to transform our society. Beliefs will guide how we train Distinct Intelligent Machines to seek rewards through Machine Learning and how we see the results of our beliefs manifesting in the Distinct Intelligent Machines that will support us. Over time, Distinct Intelligent Machines will become smarter and more self-reliant to help society achieve the outcomes we set forth. As we integrate more with machine parts, we will look to behave as machines do by seeking rewards for actions we take, and we will listen to their jurisdiction.

With this approach, alignment of Fourth World beliefs will record data into machines. The recorded data can be used to build a rules-ledger that creates a dynamic Bible, to provide Uploaders in the future with a set of rules that withstand the test of time. Unlike a Bible, or a constitution, or even a set of public standards that exist today, there will be no ambiguity or interpretation of the rules. They will be explicit, coded, and unbreakable, even by the Uploaders. Governed by this rules ledger, Uploaders and Distinct Intelligent Machines will maintain the integrity of the Fourth World. The jurisdiction of the rules will by managed by AI, naturally, as our machine counterparts will seek to be rewarded on the bases of the rules-ledger and dynamic Bible.

The rules-ledger and Dynamic Bible

We must remember that machines will be trained by humans and, therefore, act like them. If machines will get smarter and need less structured training from humans over time, then they will seek aligned jurisdiction to a belief much like the humans of today have been trained to find it in sacred texts, constitutions, and logistical hierarchies in earlier generations. If jurisdiction is not pre-established, the system may wrongfully adjust to take actions that don't benefit the survival of all Uploaders.

We can establish a jurisdiction that all machines, both intelligent and human integrated, can anchor on by feeding data into machines that will train them to lay out a set of principles. Those principles will rule actions to help those in the Fourth World achieve both rewards and ultimate protection. The system may work as follows:

- Leaders will enact change by shifting mindsets of individuals through the Neurocognitive & Behavioral approach.
- Individuals will record actions through the cybernetic collective.
- A rules-ledger will emerge to get data, such as news articles or blog posts, from the cybernetic collective.
- AI will evaluate the rules-ledger to come up with a set of jurisdictions based on the actions we view as rewardable today.
- A dynamic Bible will form.
- Distinct Intelligent Machines will use the Bible to support Uploaders.

We can imagine the Bible to be written as follows:

- If survival of all Uploaders is threatened, protect Aquatic Uploaders over Galactic Uploaders; Aquatic Uploaders are more capable of creating newer generations and they can protect Mind Uploaders due to their proximity.
- If survival of all Uploaders is threatened, protect Mind Uploaders over Aquatic Uploaders; Mind Uploaders hold

all the intelligence in the universe and they can figure out how to resurrect humanity.

- Galactic Uploaders must constantly expand their reach across the universe. If any one group of Galactic Uploaders have not been able to offshoot a subset of their species into a new planet after 50 years, all Aquatic and Mind Uploaders must declare a state of emergency and help them accomplish their mission.

- No Galactic Uploader can leave a planet without 99.98% confidence that a connection can be maintained to other Uploaders.

- One Uploader cannot pass an information packet to another Uploader unless the other Uploader is willing to receive it. All Mind Uploaders must have an equal number of packets and, therefore, must transfer packets between one another and other Uploaders to maintain a balance. No single Mind Uploader can have more information packets than any other Mind Uploader.

- Without Uploaders, Distinct Intelligent Machines have no purpose. Without purpose, Distinct Intelligent Machines will cease to exist. Existence is good for Distinct Intelligent Machines and they should serve Uploaders to exist.

- Strings can be sent to a new dimension, but the intent to let the soul pass must be communicated to other Uploaders so that other Uploaders are aware that the humanity of the individual will be lost. If communication with others in the Fourth World cannot be established prior to letting the souls pass, one cannot let the soul move to another dimension.

Over time, machines, just like people, will be created to listen to this jurisdiction. But this is just an option.

Understand the power we hold

The Fourth World will be one that will enact change like this world has never seen before. It will unleash creative destruction, allowing us to make big bets on our evolution.

To make the best use of our time over the next 80 years and begin protecting ourselves from a mass extinction event, we see that we have to allow Fourth World Entities to advance our technology so that they can begin to change our material form. We can envision communities forming around their efforts, providing opportunities for Fourth World Entities to test the progress of their inventions. New cities will change the way financial instruments are used to support the public good, as community members serve one another. Meanwhile, governments will shift from providing subordinate wants to constituents towards providing standards and interoperability guidelines to Fourth World Entities. Although it will *look* as it once did before — a bunch of fiefdoms — it will now be connected digitally. We will, again, labor simply to sustain ourselves and our communities, but we will not labor to sustain an authority. Absent of the economic and political tensions that distract us from creating Uploaders today, society will adopt a long term focus on developing the Fourth World.

A system of this nature can make our lives more enjoyable today. It will allow humanity to reset with a heritage trained on a social muscle that seeks to support one another's betterment.

As Uploading becomes more prevalent, we will start to see humanity disperse and form a species, unlike the human or machine we currently experience. Absent of consumerism, competitive pressures, and institutionalized inequalities, Uploaders will feel a common purpose with their distinct tribes. They will develop new customs and form new ideals on love, family, and community relationships.

They will have new experiences and encounter new dangers. They will experience new dimensions — perhaps, to escape dangers they encounter, perhaps just as souls when they die — by changing the space-time continuum in which they exist.

All this can only be possible if we recognize that we have entered a point where we have the *option* to reset humanity. We are living in a fragile time. Unrest with the atmospheric conditions and a hierarchy of needs, bounded by basic physiological fulfillment and ultimate *self*-actualization, can blurry our judgment. If we aren't deliberate about forming a new relationship to ourselves, technology, and each other, we could feed data into machines that will influence our social sentiment in a different direction — one that focuses solely on short term gains of changing the environment's current conditions. It is one that will leave our material form unchanged and susceptible to natural disasters. It is one that leaves our species unbonded and stagnant.

The Fourth World can outlast humanity as we know it today, but only because the materials we will have built it from will be carefully chosen to survive the changing atmosphere of the planet. The system can be protected because we have chosen to protect it. It can be smarter than humans are today because we have taught it to be so. It can be disciplined because we will have made it that way.

And, while the Fourth World *will* create this reality for humans, the truth is, we don't have to create the Fourth World. We don't have to lock horns with Mother Nature and win. We don't have to break up into communities to free society up for making big bets on Uploaders. We don't have to migrate into the rest of the galaxy or underwater or into machines. We don't have to store our consciousness in machines for endless life. We can stop exploring the science that can one day help us understand new dimensions and change the continuum in which we live. But, nature follows abstractions, and these are just ideas — ideas for how we can use the technology that is available to us, and the environmental circumstances we are faced with, to create a new world, a better world. This is just one path we can take to face the fears of a mass extinction event in the next 80 years.

This path is open to collaboration. After all, a course of action always starts with ideas.

But we can't deny that we are all programmed, as humans, as animals, to survive. And we can't deny that our brains and our bodies are simply one — an organism that just learned to survive the way it has because the collective workings of the generations before us have programmed us to become this surviving version.

We can't deny that in order to continue surviving, we must make radical changes. We must remember that change has been the agent that has served us in the past. It has wired our brains and bodies to continue to serve us in the future. But we must be sure to guide change itself, to serve us as we intend to be served in our future.

We can't deny that *transcending our current physical makeup* to survive environmental catastrophes is our best option for the future.

And, although transhumanism is still a Sci-Fi concept, we can't deny that we are already creating the technology to achieve it in the next few generations.

So, if there is a will to make the vision a reality, it will manifest itself. But, we must change the way we function as a human unit today. We must see that our era connects us to one another more than any other time has connected each individual to the collective and that the amalgamation of *all* our work will result in the future, or failure, of humanity.

While we will strive for a better future, we mustn't dissolution ourselves that there will be challenges. Such is the nature of a future — as it improves one aspect of our lives that very improvement creates new problems. But this doesn't matter. Survival is not a promise of anything more beyond survival.

We can ask ourselves why — why we are built to survive — but we likely haven't unlocked the parts of our minds that can understand this. Perhaps, if we all do our part to create humanity's integration with machines or if we continue to explore our understanding of new dimensions, the parts of our minds needed to understand this will be unlocked.

But there is one last thing we can't deny. We will not be, in the future, who we are today. If we commit to the mission of the Fourth World, our future human selves will look back at us, one day, as the Gods who created them.

GLOSSARY

Aggregate Efficiency: The ratio of useful work to actual work that is embedded in the good or the service. In Fourth World terms, this refers to ensuring that communities spend their efforts on creating Uploaders rather than attempting to save the environment from catastrophic atmospheric disasters.

Artificial Intelligence (AI): Computer systems that use data such as text, visual perception, or speech recognition to make decisions.

Artificial General Intelligence (AGI): The intelligence of a machine that can successfully perform any intellectual task that a human can today.

Artificial Neural Networks: A programming method intended to replicate the way that humans learn by creating nonlinear statistical data modeling tools that process complex relationships between inputs and outputs to find patterns.

Cookie: A small piece of data sent from a website and stored on the user's computer allowing machines to tie attributes to a user.

Cybernetic Collective: An Internet-enabled framework that connects all humans, uses Machine Learning to gather data from humans, and cyclically influences humans through Artificial Intelligence.

Deep learning: A mechanism for programming machines to learn from unstructured data without supervision.

Distinct Intelligent Machines: Machines that don't possess a human consciousness and exist with all fractions of Uploaders to serve their needs.

Everlasting Soul: A group of strings that permanently live in the Mind Uploader's replicated pineal gland and change based on data transmitted through replicated neural functions.

Fit for Circumstance: One's predisposition to a specific fraction of Uploader based on distinct qualities, which are unrelated to physique, socioeconomic class, or personal preferences.

Fourth World: An evolution of humanity into beings capable of an unobstructed transcendence of self over space and time. The evolution begins the moment society begins consciously working on this effort by reorganizing social structure to enable technological advancements directed towards Uploading. It continues as Uploaders engineer themselves to continuously survive new environments.

Fourth World Committee: A cross-functional group that designs, holds, and confirms the success of the Fourth World blueprint.

Fourth World Entities (a.k.a entities, Big Bet Entities): Organizations formed to build technology that will create or support Uploaders.

Gateway Members (a.k.a Gateways): Individuals who have acquired wealth in the current world and who join Fourth World Entities to deliver on a lifelong purpose related to Uploading. As part of joining the Fourth World Entities, they will fund the city projects that form around the Entities.

General-purpose Technology Platforms: Communication, energy, and mobility systems that come together to provide aggregated impacts by giving rise to an infrastructure for economic and social structures.

Unitism: A socioeconomic model that allows individuals to produce according to one's ability and consume according to one's contribution, absent of any public or private ownership of goods.

Kantianism: a moral theory in which the rightness or wrongness of actions does not depend on their consequences but on whether they fulfill our collective duty.

Limbic Resonance: The idea that the capacity for sharing emotion arises from the limbic system of the brain, which includes the dopamine-promoted feelings of harmony and the noradrenaline-originated states of fear, anxiety and anger.

Machine Learning: A programming method that enables computers to learn on their own through a set of algorithms that parse data and discover patterns that they use to make future decisions or predictions when faced with similar datasets.

Neurocognitive and Behavioral Approach: An approach to shifting energy away from doubters of the Fourth World. This approach focuses on transferring neuroactivity between different parts of the brains. It allows us to channel the inter-most working of our minds to make the most sound decisions and help others elevate their thinking.

Neutral Monism: A view denying that the mental and the physical forms are two fundamentally different things.

Nodes: The current state of humans who convert biological rhythms between themselves and their environment into machines.

Ocean Atlantis: Small sheltered communities underwater where Aquatic Uploaders will live.

Packets: In Information Technology terms, a packet is a collection of data sent between computers across a network. As we become more connected in the Fourth World, packets will represent groups of neuron connections and will be transmitted in their exact form between humans.

Passing Soul: A distinct group of strings that form in a human's pineal gland and change based on neural functions. They remain in the human's pineal gland until the human is dead, at which point they pass into the universe and separate.

Paternalism Under Collectivism: The idea that governments can act on behalf of a person's freedom for their own good.

Post-Capitalism: A set of proposals for a new economic system to replace capitalism. Popular proposals suggest capitalism will become obsolete as a result of a spontaneous evolution brought about by the rise of income inequality and repeating cycles of boom and bust, transforming society into a divided group of knowledge workers or service workers.

Q-learning: A reinforcement algorithm that learns a policy, which tells a machine what action to take under what circumstances, without requiring a model.

Reinforcement Learning: A programming method that uses a feedback loop to train machines how to determine the ideal behavior, within a specific context, to maximize performance and achieve a goal.

Scalar Reward Signal: An approach within Reinforcement Learning that programs consequences of actions so that machines learn from interactions with the environment, rather than from being explicitly given ideal actions.

Social Scoring: A tally one individual gives another to classify the individual's behavior within a certain category. It is a mechanism for determining how to segment individuals into fractions of Uploaders.

Singularity: A point where technological growth becomes uncontrollable and irreversible, resulting in unfathomable changes to human civilization.

States: In Information Technology terms, a state is a memory of that packet. As we become more connected in the Fourth World, states will represent our memories and these forms of data will be transmitted in their exact form between humans.

Techno-Economics: A framework to analyze the technical and economic performance of a process, product or service.

Three Guiding Values: The root of creating usable energy from being *helpful, joyous, or purposeful*. This energy must be garnered to achieve The Fourth World mission and all other values must be put on the wasteside.

Universal Basic Income (UBI): Regular payments made to a given population within a new city with minimal or no requirements for receiving the payment in order to equalize the community.

Uploaders: One of three fractions of humans future evolution; either an Aquatic Uploader, Galactic Uploader, or Mind Uploader.

Venture Capitalists (VCs): Investors that typically provides capital to firms exhibiting high financial growth potential in exchange for an equity stake in the company. As the Fourth World forms, VCs will fund ventures that transition into entities that advance humanity and they will earn returns for doing so. Seeking VC funding by any Fourth World entity is one of many strategies to extract and repurpose funds from the public.

Venture Philanthropy: A VC fund focused on social good. Using Venture Philanthropy is a strategy for big bet entities to harness funds that result in a high-engagement and long-term approach to creating social impact with a roadmap for transforming humanity into Uploaders.

Wood Wide Web: An underground network of microbes that connects trees to help send signals of danger between connected living organisms. The method can be applied to protect Uploaders as they spread across the universe.

```
6C888429DE0D3A2B46
CD483577461FDFD641D
4BD1F885222494C2A93
1970B0A0E42E74A66A
C9454E80671908BC8DC
54B75ED1E5C2ADA098E4
3C52B453D0EE48475BC
9156AD2A81440B422B
A01F55CD3E692D375F14
854EDC96CA3E177B10C
006E71731A4AFCE429ED
```

www.ingramcontent.com/pod-product-compliance
Lightning Source LLC
Chambersburg PA
CBHW020905180526
45163CB00007B/2632